LASER
LIGHT OF A MILLION USES

LASER
LIGHT OF A MILLION USES

JEFF HECHT
AND
DICK TERESI

DOVER PUBLICATIONS, INC.
Mineola, New York

To Lois, Leah, and Jolyn, who put up with everything—J.H.

Published in Canada by General Publishing Company, Ltd., 30 Lesmill Road, Don Mills, Toronto, Ontario.
Published in the United Kingdom by Constable and Company, Ltd., 3 The Lanchesters, 162–164 Fulham Palace Road, London W6 9ER.

Bibliographical Note

This Dover edition, first published in 1998, is an unabridged republication of the work originally published by Ticknor & Fields, New York, in 1982 under the title *Laser: Supertool of the 1980s.* A Preface to the Dover Edition written by Jeff Hecht has been prepared specially for the present edition.

Library of Congress Cataloging-in-Publication Data

Hecht, Jeff.
[Laser, supertool of the 1980s]
Laser, light of a million uses / Jeff Hecht and Dick Teresi.
p. cm.
Originally published: New York : Ticknor & Fields, 1982 under title: Laser, supertool of the 1980s.
ISBN 0-486-40193-6 (pbk.)
1. Lasers—Popular works. I. Teresi, Dick. II. Title.
TA1675.H43 1998
621.36'6—dc21 98–4209
CIP

Manufactured in the United States of America
Dover Publications, Inc., 31 East 2nd Street, Mineola, N.Y. 11501

CONTENTS

PREFACE TO THE DOVER EDITION

A decade and a half have passed since we wrote the first edition of this book, and the years have brought some changes to the laser world. If you're an average technophile, you probably have at least one laser in your home—hidden inside your CD player. You see lasers at work every time you go through a supermarket check-out counter. Someone you know probably has had laser surgery. You've probably seen a speaker highlight points on a screen with the bright red spot from a laser pointer.

Indeed, lasers are pervasive in our high-tech world, often hidden deep inside places we don't look. Laser printers personalize our junk mail. Tiny semiconductor lasers send signals through fiber-optic cables that cross oceans and continents, and run from one town to the next. Lasers cut elaborate patterns in greeting cards. Satellite-borne lasers probe the atmosphere, extracting a wealth of data for scientists. Laser machine tools cut plastics and weld metals.

The details of laser technology change constantly. Over the past decade, a new generation of semiconductor lasers have swept away the older generation of gas lasers, just as semiconductor electronics earlier replaced vacuum tubes. The new semiconductor lasers are powerful enough to perform minor surgery. We take it for granted that a pen-sized semiconductor laser, powered by a couple of penlight batteries, can shine a bright red spot on the wall—but that was a research breakthrough a dozen years ago. The latest breakthrough is a new semiconductor laser that shines bright blue light. As I write in 1998, those blue lasers last only a few weeks in the laboratory, but before long they may be hot techno-toys for power presentations. (One eminent laser professor pointed out that this will be a boon to the color blind who, like him, cannot see red laser spots.)

Yet for all the changes, the central ideas of laser technology remain the same. The principle that makes lasers work and gives them their name—Light Amplification by the Stimulated Emission of Radiation—hasn't changed in the forty years since it was proposed. The basic types of lasers haven't changed, although the balance has shifted to semiconductor lasers. *Laser Focus World* magazine says worldwide sales of semiconductor lasers should pass a staggering 200 million in 1997. Physicists and engineers have perfected new laser crystals and new semiconductors, but the physics remain the same. So does the way laser light affects materials, heating, writing, reading, and cutting.

The epic battle over laser patents (see chapter 4) finally ended. When the legal dust settled, Gordon Gould's lawyers had carried the day, much to the dismay of the laser industry. Companies gritted their teeth and wrote royalty checks that assured Gould wouldn't have to worry about retirement income. Gould also earned election to the National Inventors Hall of Fame, but he was not the first nor the last laser pioneer to be honored. Charles Townes had gotten there first for conceiving the principle of laser action; Theodore Maiman followed Townes for making the first working laser. Robert Hall, inventor of the first semiconductor laser, followed Gould in 1994. Maiman, who felt slighted by the Nobel Prize committee, received the more technology-oriented Japan prize for his laser work.

Laser medicine (see chapter 5) has improved greatly over the years, but it still can't perform miracles. Laser surgery couldn't save the vision in the eye of Minnesota Twins star Kirby Puckett, affected by the sudden build-up of pressure caused by glaucoma. However, another laser procedure has become the standard way to treat a common complication that can re-obstruct vision after cataract surgery. Some well-established laser treatments have also been improved. For example, precise selection of laser wavelength has greatly improved the treatment of the dark birthmarks known as port-wine stains by targeting only the abnormal blood vessels that cause the discoloration. Careful choice of wavelength also helps bleach away many—but not all—tattoo pigments.

The latest frontier for laser medicine is cosmetic surgery, some of it heavily advertised. For $1,500 to $2,000 per eye, ophthalmologists promise to correct moderate nearsightedness, but the surgery can be painful and will not avoid the need for reading glasses or bifocals in middle age. In addition, it cannot guarantee perfect vision for every patient. (I'm sticking with my glasses.) Other laser treatments are aimed at the multi-billion dollar business of making people look younger, by ironing out wrinkles and removing unwanted facial hair. On the horizon are other ideas, including a laser that zaps away the surface of the skin, leaving a thin layer that drugs can pass through—avoiding needle punctures.

Fiber optics (see chapter 6) have become the world's standard long-distance communication medium. Submarine fiber-optic cables, looking like white plastic garden hoses, but filled solid to withstand deep-sea pressures, criss-cross the world's oceans. Optical amplifiers, pieces of optical fiber doped with erbium, amplify light signals that carry billions of bits per second with sharp digital clarity. Submarine fiber-optic cables have put communication satellites out of the business of relaying transcontinental telephone conversations, although satellites still carry many other signals. However, fibers have yet to come to our homes. Today, they typically stop a few blocks away, where an optical interface box on a pole or in a manhole converts light pulses to electrical signals. These signals are then carried by wires to your telephone or your cable-television adapter. Both phone and cable companies are trying to avoid spending the extra money to bring fiber to your door, but it's likely to come eventually as existing cables decay.

The Pentagon still sees a bright future for laser weapons (see chapter 7), even after the former enemy, the Soviet Union, has gone out of business. Precision laser guidance of missiles remains a military mainstay, highlighted dramatically during the Persian Gulf War. Eye-zapping lasers have been in the news because of Pentagon proposals to build them (although their official targets are electro-optical sensors), and efforts by human rights groups to ban them. The same advances that make other lasers more compact and economical also make blinding lasers easier to build and carry on the battlefield, so the issue has become significant.

In the years since the first edition, Ronald Reagan's Strategic Defense Initiative has come and gone. Plans for space-based missile-zapping lasers have evaporated with most of the "Star Wars" budget. Several billion dollars failed to produce any practical hardware, but it took the end of the Cold War to stop the program. The latest missile-zapping scheme is the Airborne Laser, a Boeing 747 converted to house a massive chemical oxygen-iodine laser that is supposed to destroy ballistic missiles rising above the atmosphere. It will take a few years, and several hundred million dollars, to see if it works.

Industrial lasers (see chapter 8) quietly produce a wide range of products on assembly lines, even welding high-quality razor blades into their disposable housings. Lower-power lasers mark serial numbers and other codes as parts speed by on production lines. Computer-controlled ultraviolet laser beams solidify liquid gels, and layer-by-layer, build up three-dimensional models for engineering studies. But the laws of physics still assure that trying to slice bread with a laser yields burnt toast.

Laser measurement (see chapter 9) continues to expand, limited only by the ingenuity of engineers and the laws of physics. Because laser beams can remotely probe locations otherwise out of reach, they are widely used in environmental research and remote sensing. Laser measurements are vital in studies of the atmosphere; they help detect and monitor the polar ozone hole.

Laser fusion and uranium enrichment (see chapter 10) remain in development, and neither has contributed a watt of power to the global energy supply. Years of research with the Nova fusion laser at the Lawrence Livermore National Laboratory proved that fusion worked better with ultraviolet light, so the Department of Energy in early 1997 approved building a next-generation laser, called the National Ignition Facility because it is supposed to meet the criteria for "igniting" a fusion plasma. It will cost a billion dollars, and according to Livermore officials, its ability to simulate nuclear weapon explosions will play a vital role in assuring the safety of America's nuclear stockpile. France, with similar concerns, is building a similar laser. Low demand for enriched uranium—and the availability of surplus uranium from nuclear weapons—slowed development of laser isotope separation, but the laser process retains its appeal and efficiency.

Laser reading and writing (see chapter 11) are ubiquitous. Laser readers of bar codes have spread beyond the supermarket to many retail chains, and even to libraries, which put bar codes on books and library cards. One jail even put bar-coded bracelets on prisoners to keep track of their movements! Inexpensive laser printers are standard on most office computers. Built around the same mechanisms as office copiers, they print typeset-quality pages. As computer software bloats beyond the realm of floppy disks, laser-read CD-ROMs are becoming the standard way to distribute software. Fax machines have come and conquered, but standard models don't contain lasers. The laser industry can't win them all.

When the Museum of Holography (see chapter 12) went broke in New York, the Massachusetts Institute of Technology acquired its collection, which is on display at the MIT Museum in Cambridge, Massachusetts. It's well worth the visit if you're in the Boston area. Down the street, Polaroid makes holographic displays, and the MIT Media Laboratory works on computer-generated holograms and dreams of holographic video. The most intriguing holograms I've seen lately are on candy, the creation of Eric Begleiter, a Media Lab graduate and founder of Dimensional Foods in Boston. He's made a series of four lollipops featuring characters from *Star Trek: The Next Generation,* which are more fun than the holograms we take for granted on our credit cards.

Many of us have seen laser light shows at least once. But the laser's greatest entertainment success (see chapter 13) is the audio Compact Disc. The CD has made the venerable vinyl phonograph record obsolete, and become America's favorite music medium. Laser videodiscs are still around, but only among a limited number of videophiles; videocassette recorders seized the mass market first. RCA's laserless videodisc died a painful and expensive death, becoming a $500 million debacle that left the company too weak to survive on its own. An international consortium of electronics companies has set standards for a new generation of laser disks that will pack data closer together, squeezing a full feature film onto a "DVD" disk about the same size as a CD. They should hit the stores about the same time this edition is published; stay tuned.

What about the future (see chapter 14)? Inevitably, some ideas have worked better than others. Laser cooling has done wonders, bringing little clumps of atoms to within about a millionth of a degree of absolute zero. It earned Steven Chu, Claude Cohen-Tannoudji, and William Phillips the 1997 Nobel Prize in physics. Add a little extra boost from magnetic traps and tricks, and you can push the atoms into an entirely new state of matter—a Bose-Einstein condensate, where the atoms themselves are coherent, and move about collectively in the same quantum state. MIT took that trick a step further to form a series of droplets they call an "atom laser" because the atoms are coherent. (They're not really lasers because there's no amplification of the atoms, but that's a minor quibble.) A giant laser interferometer to look for gravity waves is in the works. And I've lately been talking with Stuart Kingsley, an optical physicist who has nearly completed an automated observatory to look for laser signals from extraterrestrial intelligence. He figures that if we haven't heard a word from E.T. in the radio spectrum over the past three decades, it may be because the aliens were smart enough to pick up a laser instead. It's definitely worth a look.

Jeff Hecht, March 1998

PREFACE TO THE NEW EDITION

The laser world has been a busy place since the first edition of this book was published. President Reagan put lasers into the headlines when he suggested using them to defend against a nuclear attack. Laser communications systems that transmit signals through optical fibers have spread across the country, and in just a half-dozen years have reached what some call a third generation of technology. The latest expensive toy for affluent audiophiles — the digital Compact Disc player — uses a laser to play back music with superb fidelity. Growth has occurred throughout the laser industry, and laser stocks have become hot items on Wall Street. There is even a mutual fund that invests about half its assets in laser stocks.

There have been enough changes that, if we were starting this book anew, some details would differ. Yet in looking back, we find that we anticipated most of the important developments. Some of the high-capacity laser communications systems we mentioned in chapter 6 are here already; others are still to come. President Reagan's March 1983 "Star Wars" speech would not have surprised you if you had read chapter 7 a year earlier. And the coming of the Compact Disc was foretold at the end of chapter 13, long before it was on the market. Nonetheless, we do need to touch on the high points of the last couple of years.

The biggest advances in lasers have come at the highest and lowest powers. More free-electron lasers have been demonstrated, and the prospects for taming those strange beasts to produce high-power beams look better than ever. There are unofficial reports of progress in the government's super-secret effort to build high-powered X-ray lasers, although controlling such lasers won't be easy — their energy source is a nuclear explosion. Great strides have been made in improving low-power semiconductor lasers, which are finding many new uses in communications, measurement, and in the reading and writing of information.

The epic battle over credit for inventing the laser continues, starring Gordon Gould's laser patents (see chapter 4) and a cast of a thousand lawyers. Passage of a law allowing patents to be re-examined after they are issued has added still another twist to the convoluted tale. The legal wheels continue to grind slowly, and we've lost count of the number of court cases and appeals in the works. Meanwhile, on the other side of the globe, another laser pioneer is having a different interaction with the laws of his country. Nobel laureate Nikolai G. Basov was elected to the Presidium of the Supreme Soviet in late 1982, at the same time as Soviet leader Yuri Andropov.

Laser medicine continues its steady advance, although there have been no dramatic laser "miracle cures." Much exciting work has involved the use of low-power lasers to help heal wounds, to alleviate pain, or for "biostimulation." Los Angeles physician Judith Walker has found that a low-power laser beam can cause a nerve to fire — the first evidence of real physiological effects. However, another low-power laser treatment, which purports to use laser beams to smooth out wrinkles, has generated a storm of controversy. The American Society of Plastic and Reconstructive Surgeons says that laser "facelifts" are worthless, but the unorthodox physicians who use the technique claim the real problem is misleading advertising by a few medical hucksters.

Advances in laser communications have even surprised some of the professional wild-eyed optimists who try to predict the sales of new products. Companies in the United States and abroad are spending hundreds of millions of dollars on new "third generation" systems in which a single fiber can carry thousands of telephone conversations between telephone-company switching centers. In the United States and Great Britain, some of these communication links will run alongside railroad tracks. An even stranger blend of old and new technology has been suggested in Britain — running fiber-optic cables through sewers to deliver television programs to homes. (We wonder if the people planning X-rated cable television have heard of the idea.) Fiber-optic links to homes have been tested in Japan, Canada, and Europe, but only France has shown any interest in "fibering up" many homes in this decade. The Japanese have become the leaders in installing fiber-optic links in cars partly because American automakers blew their early advantage.

It is far from clear who holds the technological lead in laser weaponry — the United States or the Soviet Union. Gloom-and-doom prophecies of a fleet of orbiting Soviet battle stations remain unfulfilled, and some analysts think the United States is ahead. The possibility of antisatellite lasers is being taken very seriously — the Air Force is considering adding them to its shopping list of new weapons — although Pentagon analysts are now doubtful that the Soviets tested such a laser against an American satellite in the mid-1970s, as originally reported. Laser weapons have yet to make a big hit on the battlefield; it took the Air Force's Airborne Laser Laboratory nearly two years, from the time it started airborne tests, to bag its first air-to-air missile. The biggest controversy now surrounds antimissile laser weapons. It's easy to poke fun at the idea by pointing out that its main advocates are conservative politicians with names like "ray gun" and "wallop," but the issue is a deadly serious one (see chapter 7 for details on Senator Malcolm Wallop's laser battle station plan). Advances in X-ray and free-electron lasers have not overcome the real limitations on laser-weapon concepts, but neither have harsh words from skeptics succeeded in consigning laser weapons to the Pentagon's boondoggle archives.

Energy may have dropped from the headlines for a while, but laser energy research continues, though with little spectacular to report. The big fusion lasers at the Lawrence Livermore National Laboratory and the Los Alamos National Laboratory have been replaced by somewhat bigger ones, and physicists think that they've learned more about fusion, but each year the fusion labs have to fight for their share of the budget. The Department of Energy did pick a laser process for enriching uranium isotopes — the atomic-vapor uranium process developed at Livermore — and that program is continuing. However, some people have begun to wonder who needs all that much enriched uranium, in light of the slow growth of nuclear power.

The spread of machining and measurement lasers in industry slowed as the recession hurt sales of all sorts of machine tools, but that trend turned around with the economy. Meanwhile, laser-automated check-out counters became commonplace at supermarkets. My six-year-old daughter now takes them for granted; while playing store one day she rang up purchases by passing items over a counter and saying "beep." Of course, the laser that reads the striped codes on food packages is rarely mentioned — it wouldn't do to let the customers confuse President Reagan's missile-zapping lasers with the innocuous ones under the check-out counter. Mercifully, talking check-out counters have yet to appear in many neighborhoods. The company that makes them claims that customers like the robot voices, but we've heard that supermarket customers in Lexington, Massachusetts, told them to shut up — appropriately enough for the place where the first shots in the American Revolution were fired two centuries ago.

The laser videodisk hasn't become a big hit on the consumer market, but the technology that makes it possible has found other uses. Audiodisk players containing tiny semiconductor lasers are reading digital recordings to reproduce ultrapure music. Modified laser videodisk players are generating the pictures and action for some video games. And a refined version of the technology can be used to store billions of bits of digital data on a reflective disk the size of a phonograph record.

Progress in some laser fields has been limited by technical, commercial, or other factors. Much of the novelty of laser displays has worn off; the laserium shows continue, but some laser artists have grown frustrated with the limitations of the medium. The hand-held game that was to use a holographic display never made it to the market; Atari, the manufacturer, became preoccupied with home computers and other products. Holography makes quiet inroads in industry, but in many ways the promise of holography is as hard to grasp as the holographic image itself.

All in all, laser technology is growing ever more important. *Lasers & Applications* magazine estimates that nearly $350 million worth of lasers was sold in 1982 and that those lasers were incorporated into systems selling for a total of $2.6 billion. The United States government is spending about three quarters of a billion dollars a year on laser research, mostly for weapons, fusion, and isotope enrichment. These figures show the extent to which the device that two decades ago was a laboratory curiosity has become a widely used tool.

What of the future? Look closely and you will see many futures, for the single label "laser" covers many diverse devices. The semiconductor lasers that promise to revolutionize telecommunications are worlds apart from the high-energy lasers that might someday revolutionize defense strategy. Lasers are already a vital part of the "information era," and their importance in the reading, writing, storage, and processing of information will grow. The laser will find a home in chemistry as it has in medicine, doing small and specialized jobs far better than anything else can. The supertool will be with us far beyond the 1980s.

Jeff Hecht, September 1983

1 INTRODUCTION: SUPERTOOL

In H. G. Wells's *The War of the Worlds,* published in 1898, extraterrestrial aliens wreak destruction on our planet with their "heat ray," a beam of energy so hot, so powerful, that it destroys anything it touches. "Suddenly there was a flash of light," wrote Wells. "It was sweeping round swiftly and steadily, this flaming death, this invisible, inevitable sword of heat." Wells's heat ray, which inspired generations of science-fiction writers to imagine ray guns and death rays, was a chillingly accurate premonition of plans to use the high-energy infrared laser as a weapon.

When real lasers finally arrived—in 1960—writers and moviemakers immediately leaped upon their destructive power and substituted the word "laser" for ray gun. A laser was one of the leading pieces of machinery in the James Bond techno-spy thriller *Goldfinger* in the early 1960s. As you may recall, Bond (Sean Connery played the role at the time) was tied down to a metal table by the villain Goldfinger, his legs spread apart, while a laser made its way directly toward his genitals. The laser's bright, red, thick beam easily cut a swath through the table. It obviously had the power to tear him asunder, lengthwise.

This is, alas, the most popular image of the laser. A ray gun. A death ray. And indeed, some lasers *do* cut through metal and some *can* be used as weapons. But this image of the laser is a reflection more of the need for drama in works of fiction than of the laser's potential usefulness in our society. Most laser beams can't cut or burn and are at best faint pencil lines in the air, light scattered by dust. The nice sharp pictures of laser beams in this book are taken by projecting the beams through clouds of smoke, which scatter their light enough to allow them to be seen and photographed.

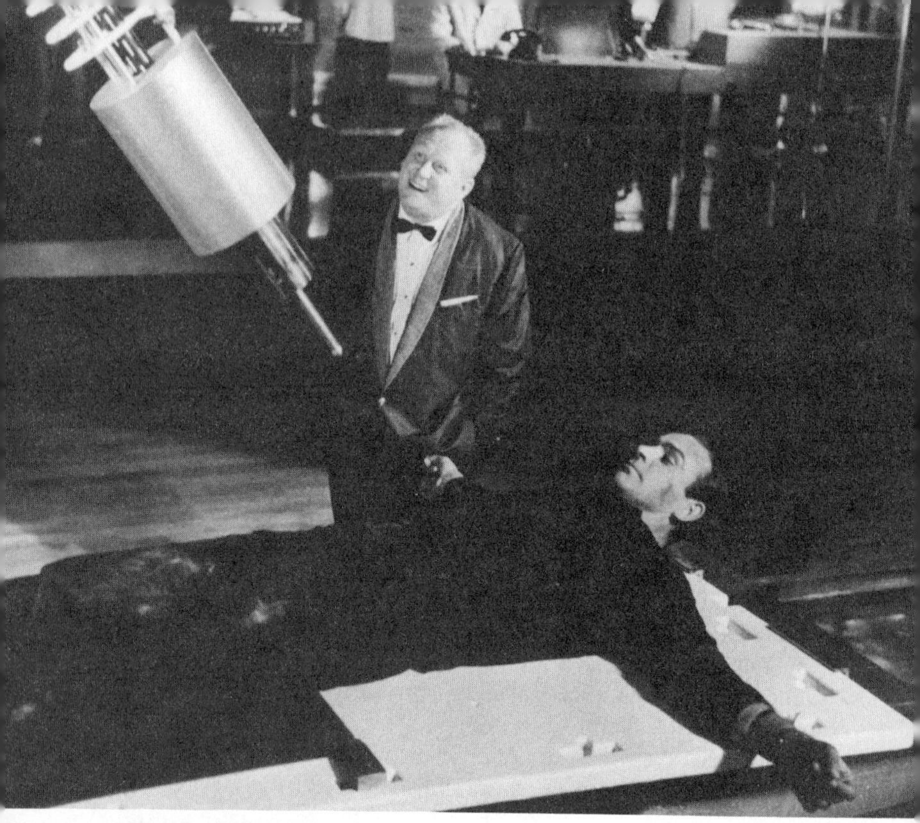

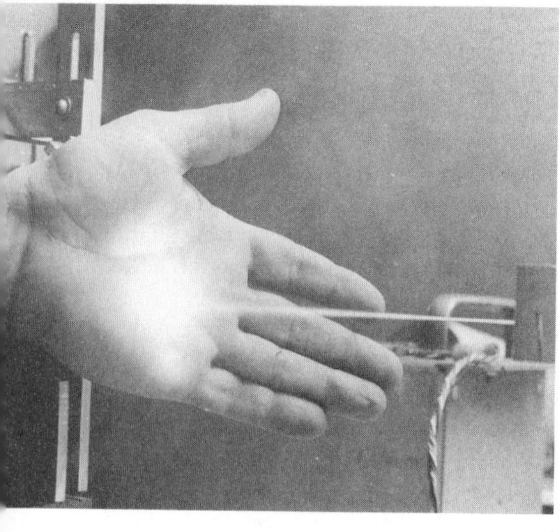

The laser, as portrayed in the movie *Goldfinger,* was simply an in strument of destruction. Courtesy Movie Star News

But many lasers are perfectly safe, such as this Associated Press Laser-photo device. The beam doesn't harm the man's hand, and in fact he can't even feel it. Encoded in the beam is information that Associated Press uses to reproduce photographs. Courtesy Wide World Photos

So how should you think of the laser? Think of it simply as a tool. One that uses light instead of mechanical energy. And a tool that allows its user to control the form and amount of energy directed at a particular place. The laser can cut through a two-inch-thick sheet of steel or detect a single atom. It can perform a task as dramatic as igniting a thermonuclear fusion reaction or as seemingly mundane as drilling a hole in a baby-bottle nipple.

WHAT IS A LASER?

A laser is a device that produces a very special kind of light. You can think of it as a super flashlight. But the beam that comes out of a laser differs from the light that comes out of a flashlight in four basic ways:

• Laser light is *intense.* Yet only a few lasers are *powerful.* That's not the contradiction you might think. Intensity is a measure of power per unit area, and even a laser that emits only a few milliwatts can produce a lot of intensity in a beam that's only a millimeter in diameter. In fact, it can produce an intensity equal to that of sunlight. An ordinary light bulb emits more light than a small laser like this, but that light spreads out all over the room. Some lasers can produce many thousands of watts continuously; others can produce trillions of watts in a pulse only a billionth of a second long.

• Laser beams are narrow and will not spread out like ordinary light beams. This quality is called *directionality.* You know that even the most powerful flashlight beam will not travel far. Aim one at the sky, and its beam seems to disappear quickly. The beam begins to spread out as soon as it leaves the flashlight, eventually dispersing so much as to be useless. On the other hand, beams from lasers with only a few watts of power were bounced off the moon, and the light was still bright enough to be seen back on the earth. One of the first laser beams shot at the moon—in 1962—spread out only two and a half miles on the lunar surface. Not bad when you consider that it had traveled a quarter of a million miles!

• Laser light is *coherent.* This means that all the light waves coming out of a laser are lined up with each other. An ordinary light source, such as a light bulb, generates light waves that start at different times and head in different directions. It's like throwing a handful of pebbles into a lake. You cause some tiny splashes and a few ripples, but that's about all. But if you take the same pebbles and throw them one by one,

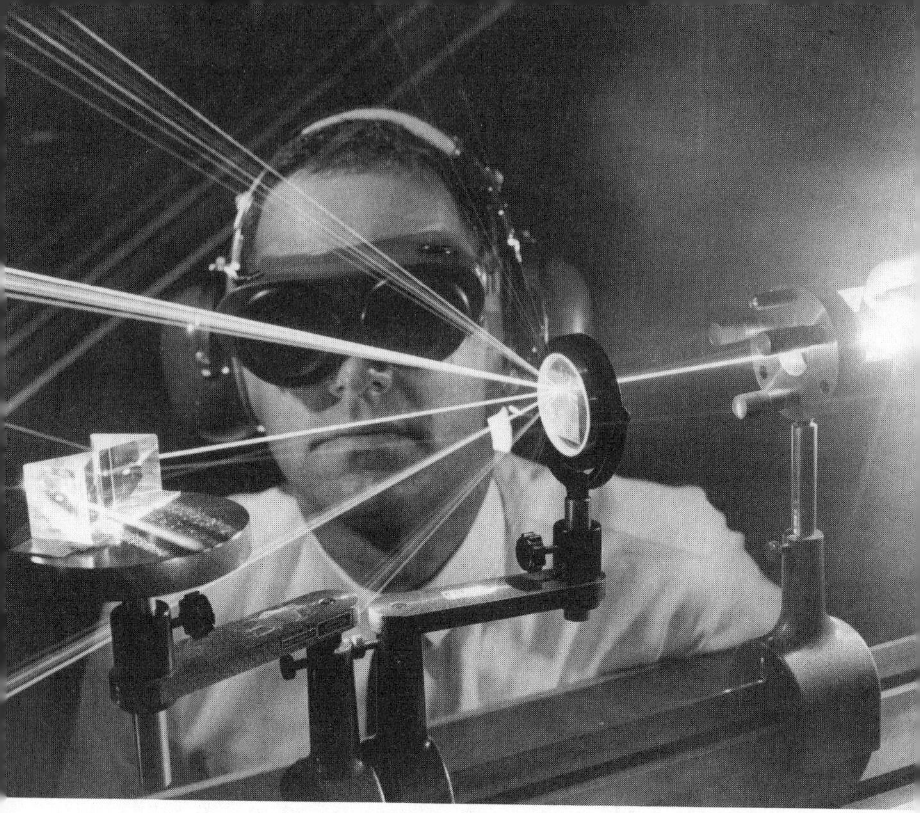

Laser beams are extremely intense and directional; that is, they are like super searchlights, able to travel long distances without spreading out or breaking up. Because of this, mirrors, prisms, and special optical elements called *beam splitters* can be used to split a laser beam into many separate beams and to bounce them all over the place. Shown here is an argon ion laser, which can be used to measure distances, carry voice and television signals, expose printing plates, and make holograms. Courtesy Hughes Aircraft Company

at exactly the right rate, at the same spot, you can generate a more sizeable wave in the water. This is what a laser does, and this special property is useful in a variety of ways. Put another way, a light bulb or a flashlight is like a shotgun; a laser is like a machine gun.

• Lasers produce light of only one color. Or, to say it in a more technical way, the light is *monochromatic.* Ordinary light combines all the colors of visible light (i.e., the *spectrum*). Mixed together, they come out white. Laser beams have been produced in every color of the rainbow (red is the most common laser color), as well as in many kinds of invisible light, but each laser can emit one color and one color only. There are such things as tunable lasers, which can be adjusted to produce several different colors, but even they can emit only one color at a time. A few lasers can emit several monochromatic wavelengths at once—but not a *continuous* spectrum containing all the colors of visible light as a light bulb does. And then there are many lasers that project invisible light, such as infrared and ultraviolet light.

We'll discuss these laser qualities—intensity, directionality, coherence, and monochromaticity—at greater length in the following chapter.

WHAT ARE LASERS GOOD FOR?

The range of uses for the laser is striking, going far beyond the original ideas of the scientists who developed the first models (though they don't like to admit this), as well as vastly beyond the visions of the early science-fiction writers, who more often than not were simply looking for a futuristic weapon (though they too are not about to admit their lack of vision).

The wide variety of lasers is also striking. At one end of the scale, there are lasers made from tiny semiconductor chips similar to those used in electronic circuits, no larger than grains of salt (a kind of laser that Gordon Gould, one of the pioneers of the field, says surprised him when it was introduced). At the other end, there are the building-size laser weapons that the military is testing, quite different from the hand-held ray guns of science-fiction writers.

Our purpose in this book is not only to explain lasers but to tell you about all the ways in which they're now used—and will be used in the near future—and about how lasers will therefore affect our lives. The tasks that lasers perform range from the mundane to the esoteric, but they usually have a common element: they are difficult or impossible with any other tool. Lasers are relatively expensive tools and are usually brought in to do a job only because they can deliver the required type and amount of energy to the desired spot. Charles H. Townes, one of the inventors of the laser and a Nobel prize-winner, told us recently that he believes the laser "is going to touch on a very great number of areas. The laser will do almost anything. But it costs. That is the only limitation."

THE $50,000 SCALPEL AND TELEVISION FIBERS

A typical surgical laser, for example, costs from $30,000 to $50,000 and up, or about a thousand times more than a good conventional scalpel. And to be honest, for many operations a scalpel may be better than a laser. But if you have a detached retina, a condition that could lead to blindness, you may be happy that these expensive scalpels exist. A laser can do what a knife can't: weld the retina back to the eyeball. No incision is required for this delicate surgery, which can be performed right in the doctor's office. The laser beam shines through the lens of the patient's eye and is focused on the retina, producing a small lesion that helps hold it to the eyeball. Exotic as this sounds, a similar laser treatment has become a standard way of curing blindness caused by diabetes. (Charles Townes finds this application of the laser amazing. He admitted to us recently that the adaptation of the laser for medical uses took him by surprise, especially the detached retina procedure.)

Laser medicine probably hasn't touched you personally (you'd know if it had), but laser communication has undoubtedly served you already. If you watched the 1980 Winter Olympics from Lake Placid, New York, or any recent football game from Tampa Bay (Florida) Stadium on television, you saw signals that were transmitted part of the way to your home by lasers. Lasers carry telephone signals in dozens of places around the country. In both cases, light from the lasers is carried through hair-thin fibers of glass—fiber optics—a technology that could ultimately bring a multitude of new communication services into your home.

DEATH RAYS, DRILLS, NUCLEAR FUSION

Lasers are already commonplace items among our military, but probably not in the way you think. Their main function in the business of war is that of range finder and target designator, not ray gun. Lasers measure the distances to targets or pinpoint them with a "bull's-eye," helping either guns or missiles to home in on the enemy. And yes, in an offshoot of H. G. Wells's idea, the U.S. military is also spending about $300 million a year trying to build lasers able to destroy targets ranging from helicopters to ballistic missiles and satellites. The Soviet Union has a comparable program and is believed to have already used a laser to temporarily blind the sensitive electronic "eyes" of a U.S. spy satellite.

In factories around the world, lasers are now used routinely to drill holes in diamonds, label automotive parts, and weld battery cases for heart pacemakers. Laser quality-control "inspectors" sit ever-vigilant on assembly lines, making sure that the sizes of parts do not deviate from an acceptable range.

One of the hopes for ending our energy problems is *thermonuclear fusion,* the process by which our sun generates its energy. One way of creating fusion here on earth is to heat and compress pellets containing hydrogen to the temperatures and pressures needed to fuse the nuclei of the hydrogen atoms together, creating tiny hydrogen bombs and thus generating incredible power. What can compress these pellets? Lasers, of course.

THREE-DIMENSIONAL IMAGES AND SUPER READERS

Lasers are what make *holograms* possible—those three-dimensional images that seem to float before you, suspended in space. But holography has many seemingly mundane applications as well, from testing the quality of aircraft tires to measuring heat flow to aiding in the design of such things as hair dryers.

Lasers have made new art and new entertainment forms possible, even beyond holography. Laser light shows, the best known of which is *Laserium,* have been seen by millions of people around the world. A laser is also at the heart of one type of videodisk player, that new device that plays back movies and television programs prerecorded on phonographlike disks.

Lasers can read. Those cryptic bar codes on food packages in supermarkets are read by scanning them with a laser beam. The pattern of reflected light is decoded to tell a computer in the back of the store what the label says. This not only tabulates the price on the cash register but automatically registers in the computer's inventory memory. Lasers also read special typewriter faces, so that manuscripts can be typeset automatically, without human aid.

And lasers can write. It's simple for a computer to control a laser, making it write on film, special paper, or the drum of a copying machine, for later transfer to paper. Lasers expose printing plates for newspapers and print statements for insurance companies and mutual funds.

A BILLION DOLLARS A YEAR

Lasers also serve the interests of pure scientific research. They are used in experiments that would require a book as long as this just to explain. Lasers can cause and control chemical reactions and someday might even propel rockets and aircraft. We'll cover all these things in the following pages.

The laser's catalogue of wonders is growing larger each day, as is the thriving laser industry. The market for lasers and related equipment hit $1 billion for the first time in 1980. That figure, which doesn't include sizeable efforts in the Soviet Union and China, is nearly double the 1977 total sales figure.

But the laser has a long way to go. Its potential is only just beginning to be exploited. While that $1-billion figure may sound impressive, it is far less than the annual sales of many companies you've never heard of. To pick a familiar name, the RCA Corporation alone sold $7.5-billion worth of merchandise and services in 1979.

In the chapters that follow, we'll tell you about the laser and how it's used—its promises and the problems it has to overcome. Don't be misled by our emphasis on problems; we're realists, not pessimists. We'd be misleading you if we pretended that there weren't problems. That's why all these dramatic new uses of the laser haven't changed our lives more extensively already.

The obstacles to be overcome demand sophisticated and elaborate techniques—or the simple, brilliant insight that leads to breakthroughs. Similar obstacles have been faced and overcome before. That's how we got to where we are now.

What we're about to tell you is the story of the laser: where it came from, what it's doing for us, and where it might eventually take us.

Before we begin, we'd like first to explain the organization of this book. We've arranged it so that you don't have to read every chapter but can skip around, reading only those chapters about laser applications in which you have a special interest. However, we do encourage you to read chapter 2—Lasers: What They Are, How They Work—first, as knowing how the laser works will help you understand the rest of the book. Chapters 3 and 4, on the different types of lasers and the recent history of the laser's various inventors, respectively, are informative but can be skipped by those of you who wish to get into the various applications of lasers faster.

Here then is the story of the laser—past, present, and future.

2 LASERS: WHAT THEY ARE, HOW THEY WORK

In most of the rest of this book, we will be looking at the laser simply as a very special type of light bulb, which produces a very special type of light. In this chapter, however, we're going to look *inside* the laser, to get an idea of what it is and how it works.

The word *laser* stands for Light Amplification by Stimulated Emission of Radiation. In this chapter, we will explain all of these terms (though not necessarily in that order) and tell you how they come together to make a laser. In the process, we'll cover a few centuries of scientific achievement. For the laser is a synthesis of the work of many great scientists. Illuminated in its beam are the ideas of Newton, Maxwell, Einstein, and many other scientists you are about to meet.

Let us begin, then, at the beginning.

IT STARTS WITH LIGHT (THE L OF L.A.S.E.R.)

Back in the early 1700s, in his book *Optiks,* Isaac Newton explained to the world how light behaves. He explained, for example, why we see rainbows after rain. The water droplets act as prisms, taking white light and dividing it into the colors of the spectrum. A prism, or a raindrop, said Newton, slows down some colors more than others, so that each emerges at a slightly different angle. This discovery is still important today to anyone who makes lenses or mirrors—or lasers.

Newton also cleverly deduced that since light travels in straight lines, it must be composed of streams of tiny particles. Newton was applauded by other scientists of the time for having arrived at this definition of light, but it turns out that he was only partly correct.

By the early 1800s, it was clearly established that light, while traveling essentially in straight lines, is not composed simply of little balls. Instead, it is propagated in *waves,* like the waves on a lake. And, like waves in water, waves of light have different *wavelengths* (see diagram 1), measured from one crest to another, and different heights, or *amplitudes,* measured from crest to trough. Light waves also have different *frequencies,* measured by the number of waves that pass a point during a certain amount of time, usually one second. This may seem like gratuitous information at the moment, but you will see its relevance when we talk about different kinds of lasers later on.

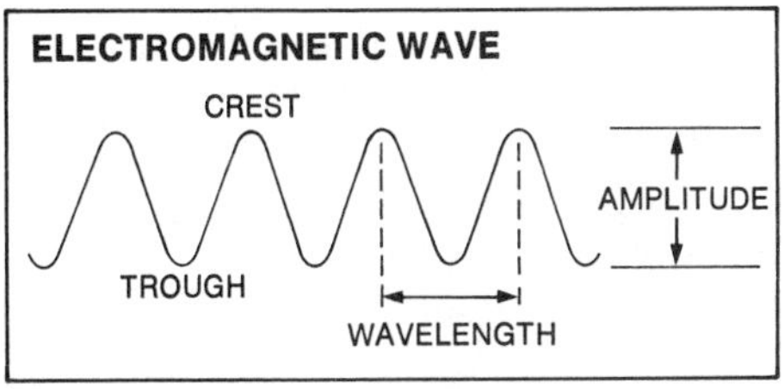

1. All electromagnetic waves, from radio waves to gamma rays, are described in terms of wavelength (from one crest to another) and amplitude (or height from crest to trough).

The wave theory of light, by the way, was proposed by a Dutchman named Christian Huygens several years before Newton expounded his particle theory. But nobody believed Huygens at the time. Well, it turns out that *both* Huygens and Newton were partly right—and partly wrong. Sometimes light acts like a wave, and sometimes, just to complicate things, it acts like a particle. We'll explain this later.

ELECTROMAGNETIC RADIATION (GIVE ME AN **R**!)

The wave theory did not explain everything about light, however. Scientists were in for some more shocks about this strange phenomenon.

In the mid-1800s, James Clerk Maxwell, a Scottish theoretical physicist who was the Einstein of his day, figured out that electromagnetic forces, which had just been discovered, travel in waves, just like light. He then discovered another coincidence: electromagnetic waves travel at the speed of light. So Maxwell made a good guess. He suggested that, just as visible light consists of a spectrum of different colors, so electromagnetic waves consist of a spectrum of different waves. And that just as the color blue is but one small segment of the spectrum of light, so all visible light is just a small part of the total electromagnetic spectrum. *Light waves are simply electromagnetic waves that you can*

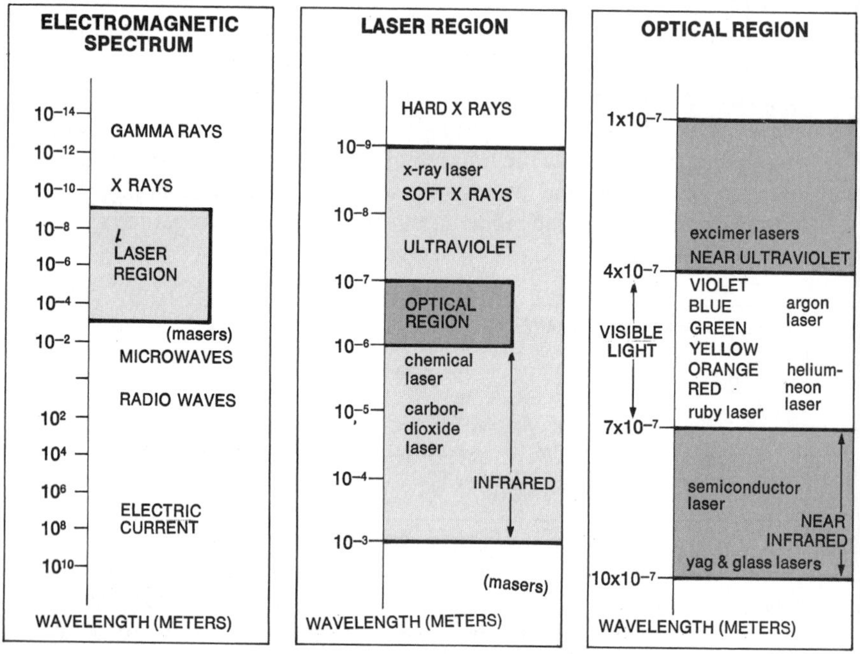

2. *Radiation* is a vague and often misused word that has come to have a sinister air associated with it in the public mind. But in physics, *electromagnetic radiation* has a very specific meaning, and it's important to your understanding of lasers to know what the term encompasses.

There are many types of electromagnetic radiation, all of which travel at the speed of light. The one with which we are most familiar is visible light, but radio waves, microwaves, infrared radiation, ultraviolet light, X rays, and gamma rays are also forms of electromagnetic radiation. Together they form the continuum shown in this chart. The longest waves, such as radio waves and electric current, are at the bottom; the shortest, gamma rays, are at the top.

This chart shows electromagnetic radiation in terms of wavelength, but you can also use the alternate scale of frequency, the number of wave peaks that passes a given point in a second. The frequency of electromagnetic radiation increases as the wavelength decreases. Quantum mechanics tells us that the energy carried by electromagnetic waves comes in chunks called photons; the energy of each photon is proportional to the frequency. This also means that the smaller the wavelength, the more energy per photon, which is why the shortest-wavelength electromagnetic radiation, gamma radiation, is the most hazardous.

The darker region of the electromagnetic spectrum (left) indicates the area in which lasers operate; this region is shown in greater detail in the laser region chart (center). Note that masers, which are sometimes considered to be a kind of laser that emits microwaves, are not really within the normal laser spectrum.

At the right is a blow-up of the optical region, which extends from the near infrared to the near ultraviolet, and in which many lasers operate. The white area in the center represents light visible to the human eye. As you can see, visible light is a very small part of the electromagnetic spectrum. And while many lasers, such as argon, helium-neon, and ruby lasers, emit visible light, there are many others that do not.

We should point out that in the chart and elsewhere in the book, we have often written the wavelengths of the various kinds of radiation using scientific notation, which is a method of writing very large or small numbers as a number between one and ten multiplied by a power of ten. For example, 5,500,000 is written as 5.5×10^6—meaning 5.5 times 10 raised to the sixth power (you can think of this as a 1 followed by six zeros), or 5.5 times 1,000,000. The same idea works for numbers smaller than zero. Thus 0.0000055 becomes 5.5×10^{-6}, which you can think of as 5.5 times a 1 that is six places to the right of the decimal point, or 0.000001.

There is also a special, standardized set of prefixes that are used in scientific measurements. Their function is similar to that of scientific notation. You're probably already familiar with *milli*meter (for one thousandth of a meter) and *kilo*meter (for units of 1,000 meters). The same prefixes can be used with any unit of measurement. We use both these prefixes and scientific notation in this book. The commonest prefixes and their values are listed below:

tera = 10^{12} (trillion)
giga = 10^9 (billion)
mega = 10^6 (million)
kilo = 10^3 (thousand)
centi = 10^{-2} (hundredth)
milli = 10^{-3} (thousandth)
micro = 10^{-6} (millionth)
nano = 10^{-9} (billionth)
pico = 10^{-12} (trillionth)

---

see, reasoned Maxwell, and there must be many other kinds of electromagnetic waves that are not visible.

Maxwell didn't live long enough to see his idea proved, but within a decade of his death, Heinrich Hertz discovered radio waves, and shortly thereafter, Wilhelm Roentgen discovered X rays. Today we know that the electromagnetic spectrum comprises many types of radiation (see diagram 2). The longest waves we know of are radio waves, some of which are several miles (or kilometers) long, such as those used for normal AM radio broadcasts. Then there are radio waves of shorter wavelengths, such as shortwave, television, and FM radio waves, and microwaves. Don't be fooled by the names shortwave and microwave, however; they're short only in comparison with the longer radio waves, which were the first to be produced and used. Microwaves, for example, are actually a few centimeters long.

It's quite a way from the microwave part of the electromagnetic spectrum to visible light, which has microscopic wavelengths—a blue light wave, for example, is around 0.00050, or 5×10^{-4}, millimeter (mm) long. The territory between the microwave and visible regions is occupied by millimeter and submillimeter radiation (whose names indicate their wavelengths) and infrared radiation (whose name means "beneath red"), all of which are invisible. Going beyond the visible region to even shorter wavelengths, you come first to the ultraviolet (beyond violet) region, then to X rays and gamma rays, the last of which are generally produced only in atomic nuclei and are so short that a hundred billion of them—each only 0.00000001, or 10^{-8}, mm long—add up to only a meter in length.

When we talk about a particular kind of electromagnetic wave, we can define it by using any one of three different measurements: its wavelength, its frequency, or the energy in each photon. Actually the three kinds of measurement are just different kinds of markings on the same ruler. Wavelength and frequency are dependent on one another: as the wavelength gets longer, the frequency decreases, and vice versa. Photon energy increases with increasing frequency and decreases with increasing wavelength. So the shorter the wavelength, the higher the frequency, and the higher the energy. If you think of the wave as a vibration and of the energy as increasing with the rate of vibration, you should not be too surprised to learn that the shortest electromagnetic waves are gamma waves, which are among the products of nuclear reactions.

That's the viewpoint of so-called *classical* physics. Electromagnetic radiation travels in waves, and the difference between two types of electromagnetic waves can be measured in terms of wavelength or frequency. But classical physics is only a first approximation to reality. To understand the inner workings of the laser, we have to look at what goes on inside the atom, and that means entering the more complex realm of atomic physics, quantum theory, and quantum mechanics.

QUANTA, PHOTONS, AND TRANSITIONS

Quantum theory was formulated in the late 1800s by the German physicist Max Planck. According to Planck, energy is not distributed evenly but rather comes in distinct chunks called *quanta,* much as matter comes in chunks called atoms. Radiation is energy, and so therefore is light. Thus, Planck showed that besides traveling in waves, light also comes in precise parcels of energy, or *photons.* However, a photon isn't quite the tiny ball that Newton imagined it to be—it's essentially a blob of pure electromagnetic energy and, strictly speaking, is massless, because it travels at the speed of light, and the *theory of relativity* indicates that no particle with mass can travel that fast.

Now let's drop the photon for a moment and see how quantum theory applies to atoms.

Quantum theory gave scientists a whole new way of thinking about matter and energy. An atom consists of a *nucleus* with *electrons* orbiting around it. Early in the century, Niels Bohr figured out that electrons don't just whiz about the nucleus in any which way. Thanks to Bohr, we now know that the electrons in an atom can travel only in distinct *orbits,* each of which is at a different distance from the nucleus. The electrons can move from orbit to orbit, but they must jump exactly into one of these possible orbits. Each orbit has a definite, fixed energy associated with it, and the *energy level* of an atom (a measure of how much energy it contains) depends on which orbits its electrons occupy.

The meaning of quantum theory is best understood by looking at the physicist's favorite example, the simple hydrogen atom, in which a single electron orbits a nucleus consisting of a single *proton.* For simplicity's sake, our diagram 3 shows a hydrogen atom with five possible orbits. (In theory there are many more, but in practice only the innermost orbits matter.) There is a unique *quantum number* assigned to each orbit, which, along with the energy level, increases with the distance from the nucleus. The innermost orbit has a quantum number of 1, and when it is occupied, the atom is in its lowest energy level. As you go outward, the quantum numbers, and hence the energy levels, increase. Hydrogen's single electron tends to occupy the lowest-energy, innermost orbit, and while there, the electron and the atom are said to be in the *ground state.*

To make a *quantum leap* into a higher orbit, an electron needs energy. A photon is a particularly convenient bundle of energy. But not any photon. It has to be one with just the right amount (quantum) of energy to move the electron precisely into another orbit.

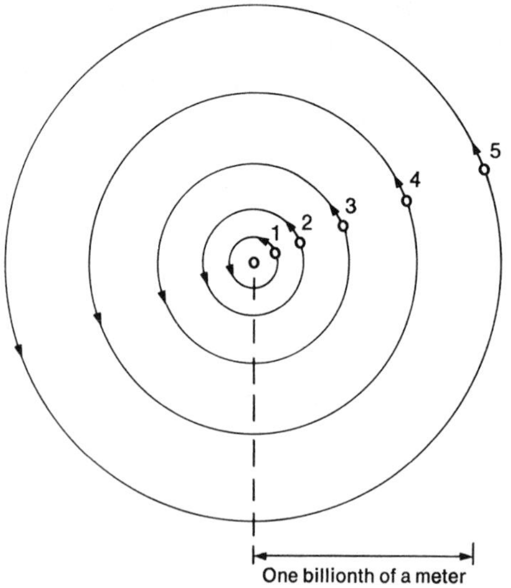

3. The simple hydrogen atom best illustrates Niels Bohr's discovery that the electrons in each atom have distinct energy levels, or orbits. At the center is the nucleus; the orbit closest to the nucleus has the lowest energy and is assigned a quantum number of 1. When hydrogen's single electron is in this orbit, the atom and electron are said to be in the ground state. As you get farther away from the nucleus, the energy levels, and quantum numbers, increase. The higher the orbit to which the electron jumps, the more energy the electron must have.

If such a photon comes along, the electron absorbs the photon and jumps into a higher orbit. The electron (and the atom) are then said to be in an *excited state.* The electron cannot remain excited for long, however, and soon—generally within a tiny fraction of a second—drops back down to its ground state. When it does so, it must get rid of its extra energy, which it does by emitting a photon, a photon of the same energy and wavelength as the one it has just absorbed. The process of moving between orbits, or energy levels, is called a *transition.* Remember this word and what it means. We will use it often when we talk about different kinds of lasers, because they are based on different transitions. We will also use it at times to identify a photon as one that has precisely the energy required for (or produced by) a transition.

THE KEY: STIMULATED EMISSION (THE S AND E OF L.A.S.E.R.)

But what happens if an electron is already in a higher orbit, in an excited state? Atoms tend to gravitate toward their ground states. So the electrons drop down spontaneously and emit photons. This process is called *spontaneous emission.*

Enter Albert Einstein. Einstein wondered what would happen if an electron were already in an excited state when it encountered a photon of the proper energy. A natural conclusion would be that the electron would move to a still higher orbit. Natural, but wrong. Einstein predicted instead that the electron would drop to a lower orbit and emit a photon, as by spontaneous emission, except that in this case the stimulating photon would not be absorbed but would simply continue on its way. We'd then end up with *two* photons of identical energy, or wavelength. Einstein called this *stimulated emission.*

This concept, formulated by Einstein in 1916, was the germ of the idea that eventually led to the laser. We have already seen that the word *laser* stands for Light Amplification by *Stimulated Emission* of Radiation. As you may have reasoned, if you could bombard atoms with the right kind of photons and catch enough electrons in excited states, you would stimulate the emission of more photons, and you'd be on your way to making laser light—that is, *a burst of light energy all at one wavelength.* The concept is similar to what happens in a nuclear reactor, or an atomic bomb, in which neutrons zap uranium atoms, splitting them apart and making them emit more neutrons, which in turn split more uranium atoms, producing a powerful chain reaction. Could the same thing be done with light?

Not too many people thought so at first. Einstein's prediction of stimulated emission was verified in laboratory experiments in the 1920s, but even in those experiments spontaneous emission was clearly dominant, and stimulated emission was rare. This is because atoms remain excited for such a short time—around a millionth of a second—that there's little chance of the right photon coming along in time to produce stimulated emission. Still, there *is* a chance. In fact, we now know that in the atmosphere of Mars, stimulated emission at infrared wavelengths occurs naturally when sunlight excites molecules of carbon dioxide. And in gas clouds near certain stars, stimulated emission in the microwave region occurs naturally. But these discoveries were not made until after the laser was invented.

PLAYING WITH MOTHER NATURE: POPULATION INVERSION

Here on earth, stimulated emission is rare and doesn't occur without human help, because nature tends strongly toward what is called *thermodynamic equilibrium,* meaning simply that atoms and molecules try to be in their lowest possible energy states. At normal temperatures, there are many more atoms in the ground state than in any excited state, and the higher the energy level, the fewer the atoms in it. This means that a photon whose energy exactly matches that of a particular transition is much more likely to encounter an atom in the transition's lower energy level (which would absorb the photon) than in the upper energy level (which could be stimulated to emit a second photon). As long as you have a so-called *normal* energy distribution, then, you can't have a laser.

Scientists realized this and knew that the only way they could produce strong stimulated emission would be to have the reverse situation: more excited atoms than unexcited atoms. This condition is known as *population inversion* and is a radical departure from normal. So radical, said Arthur Schawlow, one of the laser's inventors, that for many years physicists found the prospect unthinkable. With this mental block, it's no wonder that physicists didn't come up with the laser sooner!

To create a population inversion, you have to *pump* atoms or molecules with energy so as to move them to higher energy levels. In most cases, the atoms or molecules return to their normal states so quickly that the effort is futile. But if you pump certain atoms or molecules in just the right way with just the right energy, the result is a population inversion. The first person to do this was Charles H. Townes, though he didn't do it with light but with microwaves. Townes had worked with radar during the Second World War and had then become interested in microwaves. (Radar uses high-frequency radio waves and microwaves.) By exciting molecules of ammonia gas, Townes and his associates made the first *maser,* which stands for Microwave Amplification by Stimulated Emission of Radiation.

Ammonia molecules have two energy levels that are separated by a quantum of energy corresponding to electromagnetic waves with a wavelength of about 1.25 centimeters (cm), or roughly half an inch. Townes and his colleagues knew that both energy levels were present in normal ammonia molecules, but that most molecules were in the ground state. They didn't know how to get all the atoms into the upper level. But they did know something else: that ammonia molecules behave peculiarly when exposed to an electric field. The field attracts molecules in the ground state and repels those in the excited state. So Townes used an electric field to separate the two types of molecules. By getting rid of the ground-state molecules, he was left with a population inversion in the remaining (excited) molecules.

Once the population inversion had been produced, nature took its course. A few of the excited molecules returned spontaneously to the ground state, emitting photons at the 1.25-cm wavelength in the process. These photons stimulated emission of photons from other excited ammonia molecules, and the new photons in turn stimulated emission from yet other excited ammonia molecules. The cascade grew, unimpeded by absorption caused by molecules in the ground state.

Once the maser had been invented, scientists turned their sights to bigger game—the laser (though at the time they called it the "optical maser"). Could they find atoms or molecules that could be pumped up to higher energy levels, creating a population inversion, and then produce those special light waves by stimulated emission?

Several methods of pumping were proposed. One was called *optical pumping,* which simply means that light itself is used to excite the atoms and molecules. Today optical pumping is very common, and even sunlight has been used to pump lasers, though arc lamps, flash lamps, and other lasers are more commonplace. Beams of electrons or protons have also been used to excite lasers, as have fragments from nuclear reactions. And in some lasers, the energy comes from chemical reactions.

All of these pumping methods have been used and do work. Back in the 1950s, though, the method that most scientists, including Townes and Schawlow, thought would be the first to work was passing an electric current through a gas.

But they were wrong.

HELLO, RUBY THURSDAY

Perhaps it is fitting that a tool as glamorous as the laser was first made using a romantic crystal, ruby. Artificial ruby, to be precise; natural ruby has too many impurities.

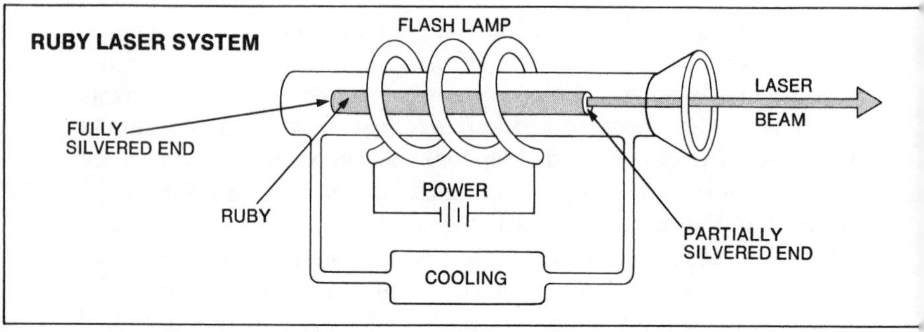

4. Theodore Maiman's original ruby laser system looks simple, but most physicists at the time believed that the first lasers would be made by passing an electric current through a gas rather than by exciting a crystal with a flash lamp. The "mirrors" on Maiman's laser were actually thin films of silver on the ends of the ruby rod. One end was fully silvered, so that it would reflect essentially all of the laser light back into the rod; the other was partially silvered, so that it would reflect some of the light back into the rod but would allow the rest to emerge as the laser beam. Cooling is needed, because only a small fraction of the light energy from the flash lamp is turned into a laser beam; the rest heats the crystal, and this heat must be removed, to avoid damaging the material.

The choice wasn't based on sentiment, however. The chromium atoms that give ruby its red color have an interesting—and useful—energy-level structure. Theodore Maiman studied ruby as a maser material for a while, and he thought that ruby would be suitable for an "optical maser." So, working alone at the Hughes Research Laboratories in Malibu, California, he started with a rod of synthetic ruby about 4 cm (1.5 in) long. Such rods of synthetic ruby have an even-color quality, almost like jellied cranberry sauce when it has been removed from the can (a bit lighter and pinker in color, however, and of course harder). Maiman then slipped a spiral flash lamp, similar to those used in high-speed stroboscopic photography, around the ruby rod (see diagram 4). When he fired the flash lamp, the chromium atoms in the ruby absorbed green and blue light. This excited a majority of the chromium atoms, creating a population inversion. The first requirement had been met.

Here it gets a bit more complex. Population inversion in the ruby laser involves three energy levels: the ground state and two excited states. The flash lamp excites the atoms to the higher of the two excited states. The atoms at this level quickly drop down, or *decay,* to the

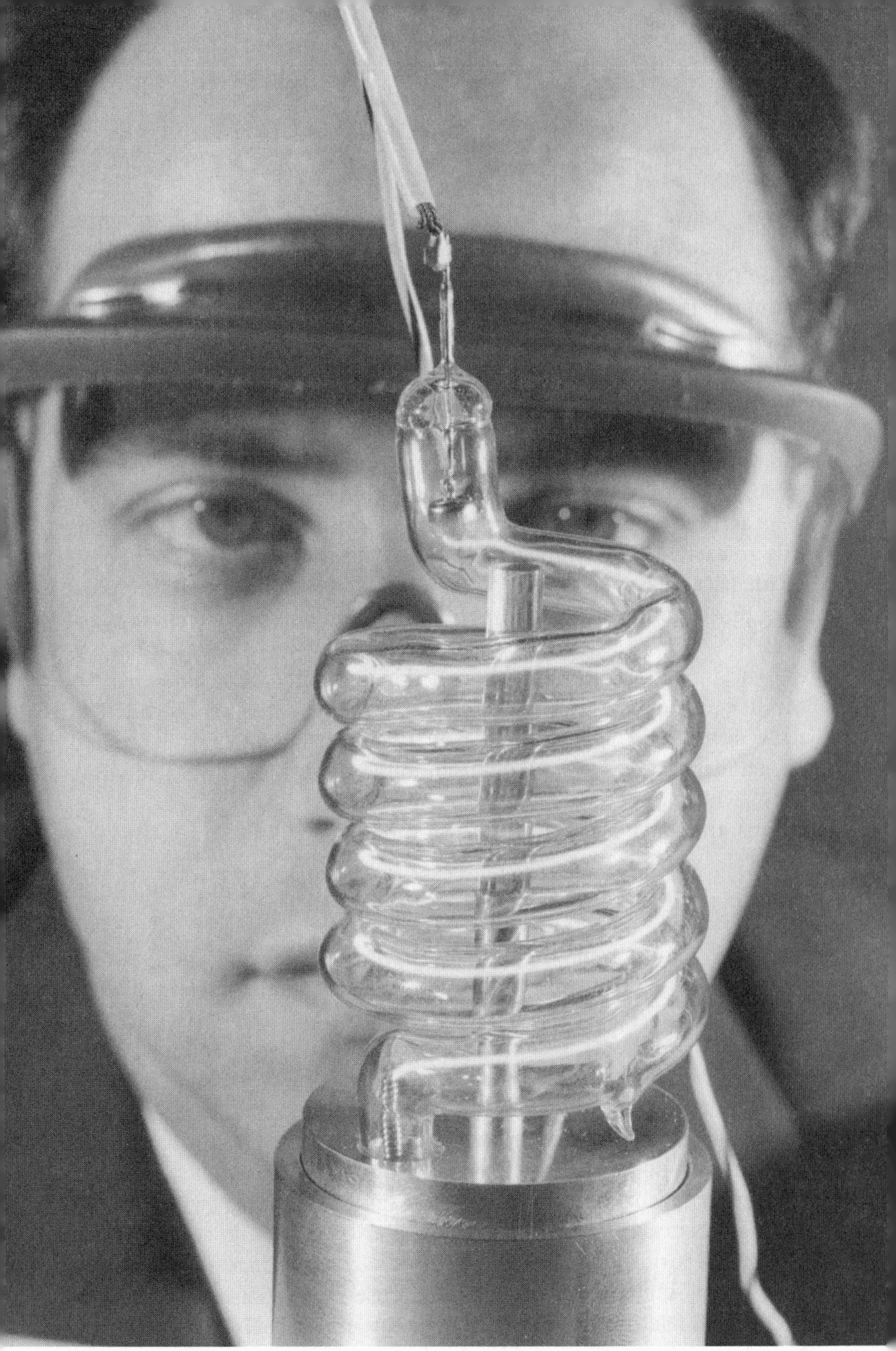

In this 1960 photograph, we see Theodore H. Maiman studying the very first laser, which he built in the Hughes Research Laboratories in Malibu, California. The spiraled translucent tube is a flash lamp, which is used to excite atoms in the synthetic ruby crystal that runs vertically through its center. Courtesy Hughes Aircraft Company lower excited level, or, as it's known technically, the *metastable state* (which sounds like cancer of some sort but isn't). Here the atoms remain trapped for what is a *long* time on the atomic scale of things, although it is only a tiny fraction of a second. While in this semistable state, the chromium atoms are still excited and can produce the long-sought stimulated emission when hit by photons of the right wavelength. Where do these photons come from?

From spontaneous emission. While in the metastable state, a few of the chromium atoms will drop back to the ground level all by themselves. As you know by now, this means that they will release a photon at exactly the right wavelength. And each time one of these photons hits an excited chromium atom in the metastable state, another photon of identical wavelength will be emitted. Now you have two free photons that go on to strike two more atoms, and you begin a chain reaction.

In Maiman's Malibu laboratory, this is exactly what happened. A bright beam of red laser light emerged from Maiman's ruby rod. The flash lasted only 300 millionths of a second, but the first laser had nonetheless been born. Maiman announced his results on Thursday, July 7, 1960.

THE SECRET OF THE BEAM: AMPLIFICATION (THE A OF L.A.S.E.R.)

It took a steady stream of ideas, beginning with Huygens and Newton, to get us to the laser. And actually, we haven't totally explained what Maiman did. We haven't talked about *amplification.*

The laser produces a thin, intense beam, yet there's nothing in the physics of stimulated emission that we've explained so far that requires that the light be in the form of a beam. Indeed, stimulated emission doesn't have to be in a beam. A photon of the right energy can stimulate emission no matter which way it's going. Left to itself, a stimulated ruby rod would simply glow red after optical pumping, without producing a beam. If that was all that a laser did, it would still be just a laboratory curiosity. What's needed is a way to produce amplification.

Maiman didn't leave the rod alone, however. He polished its ends to make them flat and smooth, then coated them with silver, so that they would reflect the red laser light back into the ruby. That made a big

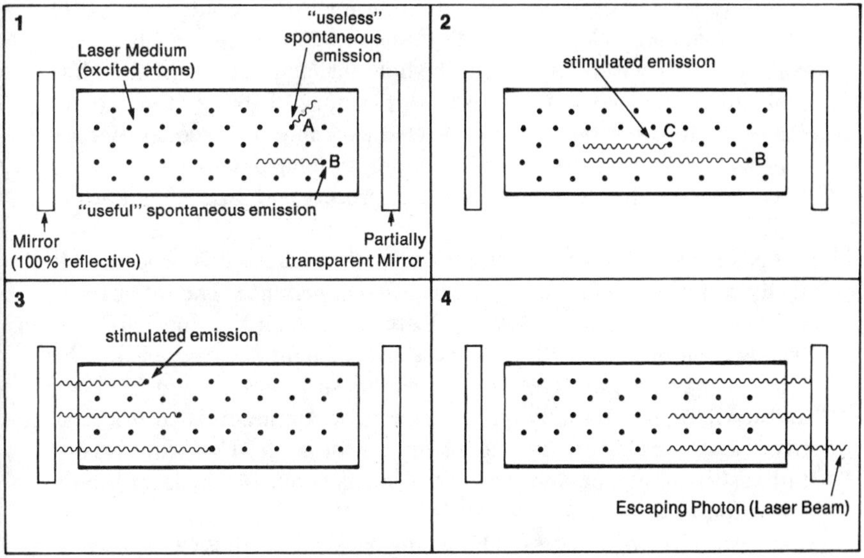

5. Here's a simplified view of what happens inside the laser cavity: (1) Atoms in the laser medium are excited, or pumped, by energy from outside (such as a flash lamp or an electric discharge). Spontaneous emission occurs naturally. The photons emitted in directions not in line with the laser mirrors, such as photon A, leave the laser medium and make no contribution to the laser beam. The photons emitted in line with the mirrors, such as B, are useful to the laser. (2) Each such photon can cause *stimulated* emission of an identical photon from another excited atom, such as C, and both photons continue to travel in the same direction. (3) These photons then stimulate emission from more excited atoms, and the stimulated light waves get bounced between the mirrors. (4) When they hit the partially transparent mirror, some photons escape from the laser cavity and thus form the laser beam. Other photons are reflected, returning to the laser medium to stimulate further emission.

difference. Stimulated emission that emerged from the sides of the rod never came back. But light that was emitted along the length of the rod was reflected back into it. The farther the light traveled through the rod, the more it grew. One photon would become two, two would become four, and so on. (In ruby, the number of stimulated-emission photons increases about 5 percent every centimeter the light travels through it.)

Obviously this sort of thing can't go on forever—it's the type of chain reaction that produces the explosion in an atomic bomb. The number of excited atoms puts one limit on the process: at some point there are simply no more chromium atoms in the excited state. But there's something else we haven't told you yet. One of the silver mirrors on the ruby rod lets some of the light through. Each time light hits that mirror, a small fraction—say 10 percent—of the photons are let out of the rod, and the rest are reflected back into it. The ones that are reflected back into the rod produce more stimulated emission; the ones that are let out are the laser beam. The process is shown schematically in diagram 5.

Let's stop a moment to retrace our path, because this process isn't really equivalent to anything in our usual experience. The intensity of light builds up inside a laser, and a certain small fraction leaks out through one mirror. Although we speak of light "bouncing" off the mirrors in a way that suggests that there's a single wave bouncing back and forth, light is actually hitting the mirror continuously over a relatively long time (on an atomic scale) in most lasers. The beam is continuously leaking out, and the beam's intensity outside the laser is only a fraction of that inside.

Technically, this process of bouncing back and forth is called *oscillation,* and it occurs in a *resonator,* or laser cavity. Townes and Schawlow, working together, outlined the basic idea of a resonator a couple of years before Maiman built his laser: you put parallel reflectors, or mirrors, on either end of a long, thin column of the substance to be stimulated. Gordon Gould also came up with the same idea, but that's a small part of a much larger controversy that we'll describe in chapter 4.

That first ruby laser is a good example of how lasers work. In the next chapter, we will look at other types of lasers: lasers that use different types of crystals or glasses, gas lasers, semiconductor lasers, and others. Before we get into specific types, though, let's take a closer look at the characteristics common to all laser light.

SUPER LIGHT

We've started with the physical fundamentals of the laser, because the laser's qualities follow logically from them. For example, the fact that a laser emits a narrow beam is a consequence of the two-mirror resonator used in almost all lasers. The photons must travel back and forth between the two mirrors in essentially parallel lines. Those that don't, leak out and don't become part of the beam.

This narrowness is what allowed a laser to be beamed at the moon and still be focused enough to bounce back a signal. The degree of parallelism is impressive. The beam from a typical inexpensive (about $100 to $500) laser will spread only about one-twentieth of a degree. That means that after traveling one kilometer (about three-fifths of a mile), the beam is only one meter (about one yard) in diameter, a spread of only one part in a thousand.

A laser beam can be focused to a much smaller point over short distances. Because the light rays in a laser are parallel, a simple lens can focus all the energy onto a spot about one-millionth of a meter (40 millionths of an inch) in diameter. This makes lasers ideal for delicate applications in surgery and for cutting or drilling various materials.

Lasers also produce light in a very narrow range of frequencies—that is to say, the light is virtually all the same color. It is often said that laser light is completely monochromatic, that lasers emit light at a single, fixed wavelength at any one time. But this is not quite true. In practice, many complex, but relatively weak, effects interact to spread the light output over a small range of wavelengths, which varies, depending on the optics and the laser material. Nevertheless, the laser can emit a narrower range of wavelengths than any other light source.

Some lasers can be made to produce multicolored light. So far we've talked about lasers that oscillate on a single transition only. But it's possible for a laser to oscillate on two or more transitions simultaneously, thereby producing light at two or more different wavelengths. Many laser light shows use such lasers to produce beams of several colors simultaneously.

WHAT IS COHERENCE?

Another of the special properties of laser light is called *coherence.* It is made possible by the stimulated-emission process. When one photon stimulates emission of another, the new photon begins life in the same *phase* as the photon that stimulated it—that is, their waves line up exactly: each crests when the other crests and hits its trough when the other does (see diagram 6). In other words, they are coherent. And they stay coherent over a large number of wavelengths, because they have the same frequency.

Here, too, we must qualify our statement. As in any real physical phenomenon, in practice there is some deviation from the ideal. The distance over which a laser is coherent—its *coherence length*—is limited to several kilometers (a few miles). In contrast, light from the most coherent nonlaser sources, such as special spectral lamps that emit nearly monochromatic light, has a coherence length of no more than a few centimeters (an inch or so), and an ordinary light bulb has a coherence length that is too short to be meaningful.

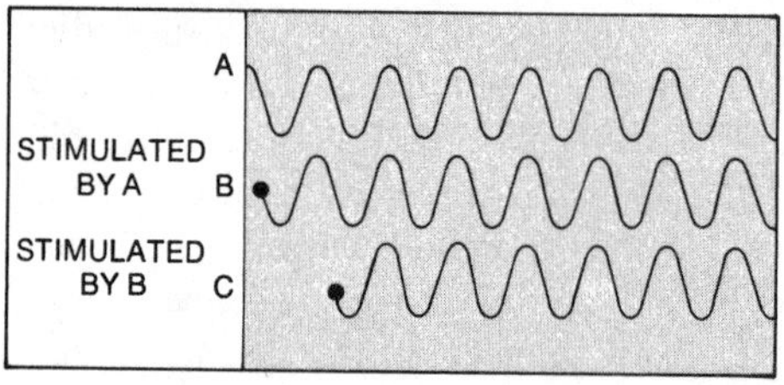

6. Laser light is coherent, because it is the result of stimulated emission. As each photon stimulates a succeeding photon, the new light wave begins life exactly in phase with the photon that stimulated it.

It is impossible to emphasize too much the importance of coherence. Coherence is a very high level of order that is exceptional in the world of macroscopic (i.e., humanly observable) physics, where individual particles are generally assumed to have random motions. It is coherence, for example, that makes possible holography, which we describe in chapter 12.

CONTINUOUS BEAMS, PULSES, AND ULTRAPULSES

Some lasers produce continuous beams; others emit short pulses. Both types have their uses. For treating a skin disease, you want a continuous steady beam, but for drilling holes in metal or trying to produce thermonuclear fusion, you need short, powerful pulses.

Some lasers can emit a beam continuously for hours, days, or even years on end. Others, like Maiman's first laser, can produce short pulses lasting a few millionths or billionths of a second. Still others can produce *ultra*short pulses lasting only a trillionth of a second or less.

The trick in producing ultrashort pulses is to keep the photons produced by stimulated emission precisely in phase as they travel back and forth between the laser mirrors. This means that all the stimulated emission photons must be the descendents of a single spontaneously emitted photon. Each time the clump of photons hits the partially transparent mirror, some of them escape to produce an ultrashort pulse. In a typical laser about one-third of a meter (about a foot) long, the round-trip time (i.e., the interval between ultrashort pulses) is a few billionths of a second. The pulses themselves are much shorter and are measured in *picoseconds* (trillionths of a second). At this writing, the record for the shortest pulse—0.09 picosecond—is held by Richard L. Fork, B. I. Green, and Charles V. Shank, physicists working at Bell Telephone Laboratories in Holmdel, New Jersey. Each such pulse is but 0.03 mm (0.01 in) long and contains only about eighty light waves.

Physicists can use these picosecond pulses as probes into chemical processes, even such mysterious things as *photosynthesis*—the process by which plants use sunlight to make carbohydrates. We'll talk about such experiments in chapter 14.

WHY LASER BEAMS DON'T LOOK LIKE THEIR PICTURES

Photographs of lasers in action almost invariably show bright beams bouncing around, but when you actually encounter a laser, you may not even see a beam in midair. One obvious reason is that many lasers emit light in portions of the spectrum, such as the infrared or ultraviolet, which are invisible to the human eye.

But you can't see the beam from even a visible-light laser unless there are enough particles in the air to scatter the light from the beam to your eye. The problem is that all the light is traveling in a straight line along the axis of the beam. If you're standing to the side of the beam, the light will reach you only if it's reflected off something. That something could be smoke, which is the trick that photographers use to make laser beams visible. It could be dust particles, which are present in air almost inevitably, and which generally scatter enough light to let you see a faint pencil line of light from a moderate-intensity laser (although often not enough to see the red beam from a small helium-neon laser). It could even be a piece of paper or some other solid object, which stops the beam. But in this case, all you will see is a round spot of light where the beam is terminated. You won't see the beam for its entire length, from the laser to the object it's illuminating.

Some people may tell you that they've seen a glow coming from some lasers. They're telling you the truth, but what they've seen is not the laser beam. Going back to laser fundamentals, remember that laser light comes from the stimulated emission of atoms of chromium or some other material. Only those light waves emitted along the axis of the laser, however, get caught up in the oscillation (the back-and-forth bouncing between the mirrors) that produces the laser beam. All the rest of the stimulated light leaks out of the laser medium and doesn't undergo further amplification. If you take the metal case off an ordinary helium-neon laser, for example, you can see a red glow coming from inside the tube. That glow is the wasted light that's emitted by excited atoms in directions other than right along the axis of the tube.

That light, by the way, doesn't just go away. Better that it would! Along with most of the energy used to excite the laser, it ends up as heat, which must be removed, lest it destroy the laser eventually. In low-power lasers, the air around us can do the job without any help. However, higher-power lasers require cooling systems using circulating water or other coolants.

The waste-heat problem is one of the most obvious manifestations of a major practical limitation of lasers: they're inefficient producers of light. In an average laser, only about 1 or 2 percent of the energy that comes in from the wall plug goes out in the beam. There is a wide range, though, and the "wall-plug" efficiency can be as high as about 30 percent or much lower than 1 percent. Inefficiency is a problem with other light sources as well. Only about 2 percent of the electrical energy that goes into an ordinary light bulb ends up as visible light, and even fluorescent bulbs turn only about 10 percent of the electricity they use into visible light.

HOW DANGEROUS ARE LASERS?

Never look into a laser beam. The lens of your eye can focus the beam to a point on the part of your body most sensitive to light, the retina, which is at the back of the eyeball. The intensity, though not the total power, of even a low-power laser beam is comparable to that of the sun, and staring into a laser beam, just like staring at the sun, could leave you with a permanent blind spot. Momentary exposure to a low-power beam, particularly one that is moving, like an accidental glance at the sun, is of no real danger, however, because the eye has an automatic aversion response.

The lasers you're likely to encounter couldn't burn a hole in you or your clothes. In fact, the beam from a low-power helium-neon laser doesn't even feel warm on the inside of your wrist. Lasers that could burn you are found only in certain military installations, laser laboratories, operating rooms, and factories that use lasers to cut or drill materials (such as aerospace plants, where lasers cut sheets of titanium for military aircraft). Such places are marked by warning signs, and all lasers that are even theoretically capable of causing damage to the eyes or other less sensitive organs are subject to stringent regulations by the Federal Bureau of Radiological Health, an arm of the Food and Drug Administration.

In fact, many people in the laser industry complain that the safety requirements are too stringent. That's typical of most regulated industries, but they do have a point. Despite some dangerous practices in the early years of laser research, and more recently by a few rock-music groups that used lasers on stage, the laser has compiled an enviable safety record. A United States government tabulation listed only about twenty accidents involving laser beams, some of which caused no permanent damage. And the only fatalities associated with lasers have been when people were electrocuted by the high-voltage power supplies used to produce electric discharges in the laser gas.

Okay, so far we've talked about those qualities common to all lasers. Now let's get down to specifics—the many types of lasers.

3 A LASER BESTIARY: DIFFERENT TYPES OF LASERS

What materials can be made into lasers? Almost any, it seems. Theodore Maiman jokes that right after scientists heard that he had fashioned a ruby rod into the first laser, nearly everybody with a crystal in his or her lab tried putting mirrors on the end of it to see if it would lase. The amazing thing is that many of these scientists succeeded.

Crystals, solids, liquids, gases—many, many different kinds of materials—are now used as the active ingredients in lasers. Laser action has been observed on literally thousands of transitions. Each transition produces a different wavelength, and these wavelengths cover a broad range, from microwaves to X rays.

The longest wavelengths, several centimeters (an inch or two) long, which are in the microwave region, are produced by masers. Masers, though discovered first, are actually regarded as a subcategory of the laser family. As their name indicates, masers operate only in the microwave spectral region. The shortest wavelengths at which laser amplification has been reported is 1.4 nanometers; that's 1.4 billionths of a meter (5.5×10^{-8} in), at the long-wavelength end of the X-ray region of the spectrum. This feat was accomplished by a group at the Lawrence Livermore National Laboratory, according to a report in *Aviation Week and Space Technology* that has yet to be confirmed in the scientific literature—or by Livermore, where the official reaction is "no comment" because of possible military applications.

If you take a careful look at the last paragraph, you'll note that the family of laser devices covers a wavelength range of over a millionfold. Visible light is only a tiny part of the range of laser wavelengths, just as it is only a tiny part of the electromagnetic spectrum as a whole, and you should realize that most lasers emit beams of light that are not visible to the human eye. Starting with the longest wavelengths, lasers emit microwaves, millimeter waves, submillimeter waves, infrared light, visible light, ultraviolet light, and, apparently, X rays. These lasers have many things in common, but there are also some striking differences. In the submillimeter region, for example, a wire mesh may serve as a mirror, despite the fact that it obviously can't be used as a mirror in the visible region.

As we describe various types of lasers, we'll try not to overwhelm you with numbers. But there is one number that's important: that's the wavelength. The wavelength of a laser defines the kind of radiation it emits and also identifies the type of laser. Remember that wavelength, frequency, and photon energy are all interchangeable terms that can refer to the same lightwave. And remember, too, that each type of laser emits light of a unique wavelength.

Besides having different wavelengths, there are also big differences among lasers in power and operating conditions. Some lasers are feeble things, which at best produce pulses of a millionth of a watt lasting only a millionth of a second. Then there are huge systems producing over a million watts for a few seconds at a time. Some lasers work only when cooled to extremely low temperatures, 200 degrees below zero on the Centigrade scale. Others need temperatures high enough to vaporize metals, because the laser wavelength is produced by a transition in a metal vapor.

In an effort to bring order out of this chaos, we will categorize lasers by the kind of materials that produce the light. The first four categories—crystal (and glass), gas, semiconductor, and liquid lasers—are straightforward. The last two—free-electron and X-ray lasers—are the strangest members of the laser family.

CRYSTAL AND GLASS LASERS

Theodore Maiman made that first historical laser from a crystal of synthetic ruby, and *ruby lasers* are still in common use today and are typical of the crystal/glass class of lasers.

Synthetic ruby, like the natural gem, is aluminum oxide that contains a small percentage of chromium as an impurity. This impurity, the chromium, is what gives the crystal its red or pink color and is the

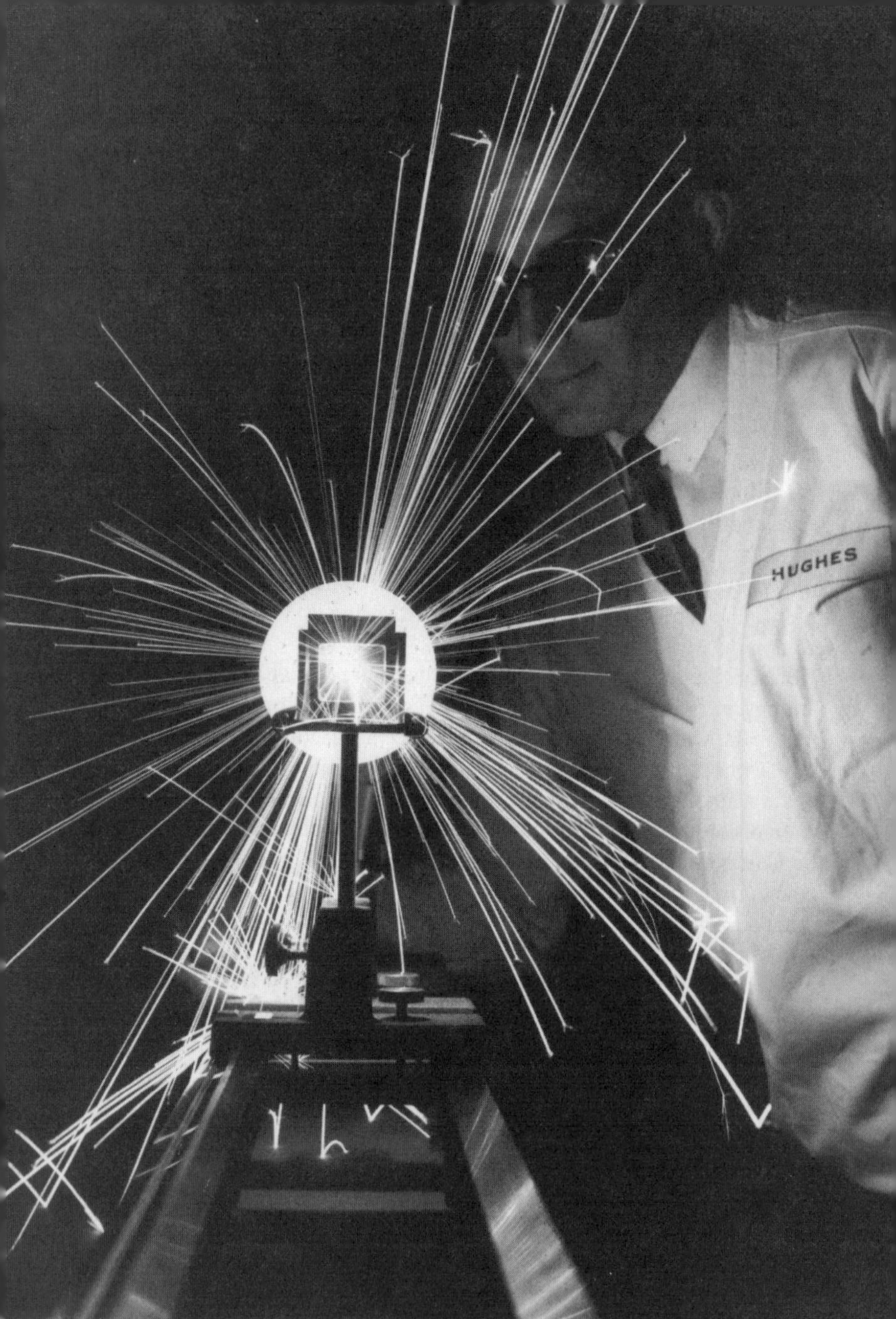

A ruby laser pierces a hole through a sheet of hard tantalum metal. When the laser was first invented, researchers liked to experiment with their new "toy" by seeing how many razor blades each pulse from the laser could penetrate. Courtesy Hughes Aircraft Company active material in the laser. The aluminum oxide serves merely as the "host" crystal, the lattice in which the chromium atoms are embedded.

The red color in ruby comes from the fluorescence produced by the chromium atoms after they absorb light. To be precise, chromium selectively absorbs light waves of 550 nanometers, or 5.5×10^{-7} m. As we saw earlier, chromium then drops to a metastable energy level, which in turn produces visible red light at 694 nanometers when it decays to the ground state. (The longer the wavelength, the less energy the light will have; that's why the waves get longer when the chromium drops to lower energy levels.)

A ruby laser can be made large enough and powerful enough to drill small holes in thin sheets of metal. Early laser researchers liked to punch holes in razor blades, and so they began measuring laser power in "gillettes." A 3-gillette laser, for instance, could drill through three razor blades in a row. This tongue-in-cheek way of measuring has since succumbed to a more technical way. In fact, the separate insertable razor blade has itself succumbed to more sophisticated shaving equipment. Once again, romanticism and humor lose out to advanced technology. Pity.

The ruby laser was the first to be used in materials working, such as drilling holes in the diamond dies through which certain types of wire are drawn. Unlike many other types of early lasers, the ruby laser is still commonly used today, though its applications are restricted by its limited power and the time between pulses needed for the crystal to cool.

Today the most common crystal laser is called a *YAG laser.* Another synthetic material, YAG is a member of the garnet family of crystals and is composed primarily of the elements yttrium, aluminum, and oxygen, to which a small amount of neodymium is added. The name YAG is an acronym for yttrium-aluminum garnet. Just as the ruby acts as a host for chromium atoms, so the YAG is the host for neodymium atoms, which emit a strong laser beam at 1.06 micrometers (1.06×10^{-6} m), a wavelength in the infrared region, slightly longer than that of the longest visible light. Like a crystal of ruby, a YAG crystal is pumped by a flash lamp or a similar intense light source. This crystal conducts heat well, so it can emit a continuous beam as well as pulses. YAG lasers are used for drilling holes in metal and as military range finders, among other things.

Neodymium can also be added to glass to make a *neodymium-glass laser.* Glass is substituted for YAG, because it's cheaper and easier to make, particularly when big chunks of laser material are needed. The light produced is almost identical in wavelength with that of the YAG laser. It does differ somewhat, because of the interaction of the neodymium atoms with the host, but by less than one percent. The drawback is that glass is a poorer conductor of heat than YAG and is therefore impractical in applications in which a laser must be pulsed rapidly or emit a continuous beam.

Many other crystalline lasers have been demonstrated in the laboratory, but only a handful have ever been produced commercially. The most important of these use other crystalline hosts for neodymium and chemically similar elements. None has yet come near achieving the importance of the ruby, neodymium-YAG, or neodymium-glass lasers, though. Like ruby and neodymium lasers, all are pumped by light from a flash lamp or other intense light source (sometimes another laser is even used).

GAS LASERS

Three of the earliest American laser pioneers—Charles Townes, Arthur Schawlow, and Gordon Gould—all concentrated their efforts originally on building gas lasers. Maiman beat them with the ruby, but the instincts of the three turned out to be good. Today over 5,000 laser transitions in gases are known: about 1,300 in atoms and the rest in molecules. (Some of these transitions are in ionized gases called *plasmas,* which some physicists consider to be a separate state of matter, but that's a fine distinction, which isn't important here.)

The most familiar laser of any kind is a variety of gas laser—the ubiquitous helium-neon laser, found in high-school physics laboratories; in automated checkout systems in supermarkets, where it's used to scan those funny price codes on products; and on construction sites, where it's used to align walls and buildings. Although dollar sales of crystalline lasers are only slightly lower than those of gas lasers ($105 million versus $125 million in 1980, according to estimates by *Laser Focus* magazine), many more types of gas lasers are in common use. And because gas lasers are generally less expensive, there are many more in use than crystalline lasers.

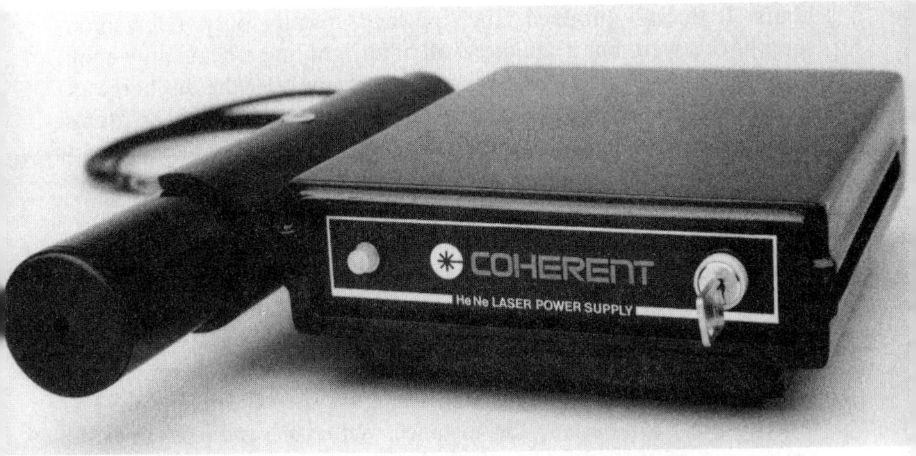

The helium-neon laser is one of the most common types of laser and is found in high-school labs throughout the land. The boxlike component of this particular model is the power supply; the beam comes out of the snap-in-and-out cylindrical head at the left. The key to the right fits the safety interlock required by federal regulations. Courtesy Coherent, Inc.

Gas lasers make up such a large category that they are further divided into families in several ways. We will take the somewhat arbitrary, but simple, approach of dividing them according to the ways in which their atoms or molecules are excited. In other words, by the ways in which they are pumped.

The most common method of pumping, or energizing, a gas laser is to pass an electrical current through the gas (generally a mixture of two or more gases) in the tube. In theory the idea is simple. You put two electrodes on opposite sides of the gas and apply a large enough voltage to get current to flow between them. This generally takes a couple of thousand volts, which is why it's important to be careful around laser power supplies. The result is what's called an *electrical discharge* in the gas: electrons flow through the gas between the two electrodes. In the process of flowing through the gas, they transfer some of their energy to the atoms in the gas. After a little interatomic juggling of energy, a population inversion is produced.

The commonest example of an electric-discharge laser is the familiar *helium-neon laser*. It can emit light continuously for many thousands of hours. It doesn't produce much power, typically only a few thousandths of a watt, but it can direct all of its light into a beam only about 1 mm (0.04 in) in diameter. In the helium-neon laser the discharge excites the helium atoms, which transfer their energy to the neon atoms, which then emit red light. Similar things happen in a red neon lamp or a fluorescent lamp, but of course these don't produce laser beams.

There are several similar types of lasers in which helium gas is excited and then transfers the excitation energy to another atom, often a metal vapor, which then emits the laser beam. One example is the *helium-cadmium laser,* which emits a blue beam. Cadmium is a soft, bluish white metal, which is vaporized into a gas by being heated to several hundred degrees Centigrade. Some of these lasers can emit at two or more different wavelengths, depending on the internal optics. The first helium-neon lasers, for instance, actually produced invisible beams in the infrared region. Only later were the red 633-nanometer lasers now in common use developed.

Other members of the normally chemically unreactive family of "inert" or "rare" gases besides helium can also be excited by electricity. Argon and krypton are the best examples, both of which can emit a whole family of wavelengths, most of which are visible. The two gases can also be mixed to get laser light at most of the visible wavelengths emitted by either gas. *Argon lasers,* which are more powerful, are generally used in industry and research. *Krypton lasers* and lasers containing both argon and krypton have a more interesting range of colors and are used mostly in light shows.

The gas lasers we've described so far all produce beams continuously but are limited in power. It's impractical to build helium-neon lasers more powerful than 0.05 watt or argon lasers more powerful than 100 watts.

Another type of laser powered by an electrical discharge is much more powerful. This is the glamorous *carbon dioxide laser.* It emits light in a band of wavelengths centered near 10 micrometers, or 0.01 mm. The wavelength of this invisible infrared light is about twenty times longer than the visible wavelengths produced by helium-neon, argon, and krypton lasers. Carbon dioxide lasers can continuously produce powers from less than a watt to hundreds of thousands of watts. The latter would be military lasers that presumably operate only a few seconds at a time and are classified. Carbon dioxide lasers can also be designed to produce extremely short pulses of even higher powers.

The carbon dioxide laser is the workhorse of the high-power industrial lasers. Here, the beam from a carbon dioxide laser welds an automotive transmission gear assembly. The curved lines are sparks flying; the infrared laser beam is invisible. Courtesy Avco Everett Research Laboratory, Inc.

The carbon dioxide laser has become the workhorse of the high-power industrial lasers. One reason is its high efficiency, up to about 30 percent, compared to about 1 percent for the neodymium crystalline laser. This high efficiency, combined with the relative ease of removing excess heat from a gas, means that the cooling problems that limit how much a crystalline or glass laser can be used are avoided.

There's a subtle, but significant, difference between the internal workings of the carbon dioxide laser and those of the other lasers we've talked about so far. The energy used to excite carbon dioxide does not raise an electron to a higher orbit; instead, it causes the atoms in the molecule to vibrate. The laser is thus said to operate on a *vibrational* transition (that is, on a transition between two vibrational energy levels) rather than an *electronic* transition (when an electron moves between two orbits). There is a practical significance to all this, because electronic transitions generally produce more energy than vibrational transitions. Thus, lasers operating on electronic transitions produce wavelengths in, or near, the visible region, whereas vibrational transitions produce longer wavelengths, in the infrared, generally at least 2 or 3 micrometers long.

Another discharge-excited high-power laser is the *carbon monoxide laser,* which operates on a vibrational transition around 5 micrometers, or 5×10^{-3} mm, in the infrared range. This laser would probably find wider use if the carbon dioxide laser weren't so good. But it has problems. Its 5-micrometer light is absorbed more strongly by air than the carbon dioxide laser light. The carbon monoxide laser is generally a more cumbersome machine as well and must often be supercooled to extremely low temperatures.

A variation on electrical-discharge excitation is the piping of beams of electrons into the laser gas. This technique is used where it's important to get a large amount of energy into the gas quickly, such as in high-power, pulsed carbon dioxide lasers.

An electron beam is, basically, the next step up in power from an electric discharge. In an electric discharge, electrons flow from a negatively charged plate to a positive plate, which attracts the negatively charged electrons. You can see the same type of thing happening in a thunderstorm: lightning. An electron beam starts out as a discharge, although generally with higher-energy electrons than in an ordinary discharge. Then it passes through a series of electromagnetic fields, which both accelerate the electrons so that they have more energy and focus them so that they're directed precisely. What this means if you're building a laser is that you can dump a lot more energy into the gas with much better control over where it goes than if you were to use an electric discharge. You can also put the energy into the gas more quickly. All these things can be very important.

They can also end up being very expensive. Electron-beam generators are large and carry a hefty price tag. They're part of a continuum of particle generators, topped by such massive systems as the 2-mile-long (3 km) linear accelerator at Stanford University. The ones used to drive lasers are much smaller than that, but they're still cumbersome and costly. One commercial electron-beam generator bears a striking resemblance in size and general exterior appearance to a dumpster. We can assure you, however, that the innards are much more sophisticated. For comparison, you can get the couple of thousand volts you need to drive a helium-neon laser from a power supply smaller than this book.

EXCIMER LASERS

We've given excimer lasers a separate heading, even though they're still in the category of gas lasers, because they're particularly interesting and important. They're also intermediate, in a sense, between the electrically driven gas lasers we've been talking about and another important type of gas lasers—those driven by chemical reactions.

In an excimer laser, energetic electrons from a beam or a discharge deposit energy in the laser gas. So far an excimer laser sounds like an ordinary electrical laser. What that energy does, however, is very different. It causes a rare gas (argon, krypton, or xenon) to react with a halogen (chlorine, fluorine, bromine, or iodine) to form an excimer, a molecule that can exist only in an electronically excited state. When the excimer emits a photon, whether by spontaneous or stimulated emission, it breaks up into its constituent atoms instead of going to its (nonexistent) ground state. This breakup automatically ensures that the population of the lower level of the laser transition remains zero, simplifying the requirements for obtaining a population inversion. If you've produced excimer molecules, you have by definition produced a population inversion. Clever!

Excimer laser are a development of the mid-1970s, which have only recently come into commercial production—even now, almost exclusively for use in research. Their characteristics are attractive for many applications, however, particularly their ability to produce high powers in the ultraviolet region, where high laser powers are few and far between. Ultraviolet lasers are valuable in photochemistry for breaking up molecules and marking silicon wafers used in electronic components.

CHEMICAL LASERS

Gas lasers can also be fired, so to speak, by chemical reactions alone. The most important of these reactions occurs in the *hydrogen fluoride laser,* in which an atom of hydrogen combines with an atom of fluorine to produce hydrogen fluoride. The combination of the two atoms produces energy, and the initial product is a molecule of hydrogen fluoride in a vibrationally excited state. Ordinarily that energy will be dissipated as heat, but if you put the excited molecules in a laser cavity and adjust everything correctly, you can extract the energy in the form of a laser beam.

In practice, the hydrogen and fluorine flow into the laser, where they are ignited to produce a flame. The flame produces excited hydrogen fluoride. Slightly downstream, the gas containing the excited hydrogen fluoride passes through a laser cavity with mirrors at each end. The cavity extracts a laser beam with a wavelength of about 3 micrometers (3×10^{-3} mm), which is in the infrared region. Fresh hydrogen fluoride passes through the cavity continuously, maintaining continuous laser action. Pulsed hydrogen fluoride lasers are also possible. One of the biggest problems in designing large chemical lasers is aerodynamical—getting the gas to flow properly—and such lasers tend to look like wind tunnels.

A prime attraction of the hydrogen fluoride laser is its ability to produce high power. What's more, you can generally store more energy per pound in the form of chemicals than in the form of electricity. And you don't need a bulky electrical power supply, such as the dumpster-size electron-beam source we mentioned earlier. All you need is something to hold the chemicals and something to ignite them. In practice, neither pure hydrogen nor pure fluorine are widely used, because they are both difficult to handle; instead laser developers use compounds that can free these elements inside the laser. It's also possible to select compounds that ignite upon contact with each other. Such lasers are most attractive for the military laser weapons programs we'll describe later, and military researchers are hard at work on many of the details.

Military researchers are also developing hydrogen fluoride lasers in which *deuterium* is substituted for normal hydrogen. Deuterium, sometimes called *heavy hydrogen,* is simply a hydrogen isotope with twice the normal mass. The nucleus of deuterium contains a proton and a neutron, whereas the nucleus of ordinary hydrogen contains only a single proton. It is used in hydrogen bombs and fusion research. Although deuterium is much more expensive than regular hydrogen, the slightly longer (4-micrometer) wavelength of *deuterium fluoride lasers* penetrates the atmosphere better, an important feature if you're shooting through the air.

OPTICALLY PUMPED GAS LASERS

Optical pumping, which we mentioned earlier in connection with crystal lasers, is also used with some gas lasers. The idea is simple and straightforward and can be used with any type of laser. All it takes is a laser or other source that produces light with a wavelength that can excite the laser medium. Unfortunately low efficiency is inherent in optical pumping, because light sources in general are inefficient. In practice, electrical excitation is usually more efficient for gas lasers. (You can't use electrical excitation for crystal or glass lasers, because the electrons can't get inside the solid.)

In general, optical pumping of a gas laser is used only when nothing else can produce a particular wavelength. That's not an uncommon situation in research laboratories, and there are a few such lasers produced commercially. Many are custom-built in laboratories. Their power levels are, almost without exception, low.

Researchers hope someday to build high-power optically pumped lasers using a light source whose efficiency is not a problem—the sun. Such lasers would be used in space for power or propulsion (see chapter 14). Although the concept has been demonstrated in the laboratory, any practical use is many years away.

NUCLEAR-PUMPED GAS LASERS

Using atomic fragments from nuclear fission reactions to excite gas lasers is an approach that was once thought to have great promise, but one that has so far yielded only modest results. The idea is to stick nuclear fuel in the laser cavity. It can either be coated on the walls or inserted as a gas. Neutrons would split the nuclei of the atoms of nuclear fuel, producing energetic fission fragments, which would then excite the laser medium, which in turn would produce a laser beam.

The National Aeronautics and Space Administration (NASA) began investigating this concept, as well as solar-pumped lasers, in an effort to find ways of transmitting power between satellites. NASA's original idea called for a nuclear reactor with a core containing the gaseous nuclear fuel, such as uranium hexafluoride, rather than solid uranium compounds. The core would be included in a laser cavity, and much of the energy produced would be extracted directly in a laser beam, without first being converted into heat and electricity, as happens in a regular reactor. The military has also been interested in such lasers, because of their predicted high-power output.

The first nuclear-pumped lasers were demonstrated in 1974, but since then progress has been slow. Experiments require special reactors that produce intense bursts of neutrons lasting only a small fraction of a second. The current record for the output of a nuclear-pumped laser is only about a thousand watts, set in 1980 by Russell DeYoung of Miami University of Ohio and his colleagues. Since then, NASA has stopped its program for lack of the money required to build the special reactor that would be needed to reach higher powers. Military interest seems almost nil.

SEMICONDUCTOR LASERS

So far we've talked about lasers that have few close relatives in our everyday lives, although fluorescent and neon lamps are vaguely similar to some gas lasers, because electric currents flow in both, exciting helium atoms, which subsequently emit light when they return to a lower energy level. But the semiconductor (or diode) laser has a much closer relative, which you've no doubt seen: the *light-emitting diode,* or *LED.* The red displays on many pocket calculators and digital watches are made up of arrays of tiny LEDs, each one a dot on the display. (The silvery displays on other types of calculators and watches are something else altogether, liquid crystals.)

To understand what happens in semiconductor lasers and LEDs, we first need to know a bit about semiconductors. Semiconductor materials, such as silicon, come by that name because they conduct electricity better than insulators but not as well as true conductors. By carefully controlling the composition of the semiconductor material, you can control how it conducts electricity—making it possible to build all kinds of useful structures, such as complex integrated electronic circuits.

The operation of LEDs is similar to that of semiconductor lasers, and their structures are similar electrically. In both, an electrical current excites carriers of positive and negative charges in the semiconductor, and these charges subsequently combine, thereby neutralizing each other. Photons are produced in the combination process, which is basically a transition from a higher to a lower electronic energy. In an LED, spontaneous emission is dominant. That's what you're seeing when you read your pocket calculator. But in semiconductor lasers, the end facets are cut to reflect light—to act as mirrors—and the operating current is higher. The result is dominance of stimulated emission.

Like other semiconductor devices, semiconductor lasers are small and can be made relatively inexpensively in large quantities. Some are the size of a grain of salt or even smaller. Although semiconductor lasers are efficient as lasers go, heat dissipation is a problem because of their small size, and power output is low. Another drawback to their smallness is their unusually large beam divergence—10 to 20 degrees or more, which is a hundred times that of inexpensive helium-neon lasers—due to the short length of the resonator. It's like the inaccuracy of a snub-nosed revolver compared to that of a long-barreled rifle. For many applications, though, a suitable focusing lens can cure this problem. Semiconductor lasers are essential to fiber-optic communications (see chapter 6).

Semiconductor-laser technology has probably been the fastest-moving area of laser technology over the past few years. The first commercial semiconductor laser able to emit a continuous beam at room temperature came on the market in 1975. Now researchers at Bell Telephone Laboratories project room-temperature lifetimes of a million hours (about 100 years) for continuously emitting semiconductor lasers made in their laboratory. Some researchers at Varian Associates in Palo Alto, California, even project "geological" lifetimes (millions of years) for some new LEDs they've developed. Other advances include development of new structures able to produce beams of better quality.

Developers have also opened up a whole new realm of materials. For many years, most semiconductor lasers were made of compounds of gallium, arsenic, and aluminum, which emit light with wavelengths between 800 and 900 nanometers, or 8×10^{-4} and 9×10^{-4} mm, in the infrared range. Recent work includes development of a new family of lasers made of indium, gallium, arsenic, and phosphorus, which emit light at wavelengths between about 1.1 and 1.6 micrometers (11×10^{-4} and 16×10^{-4} mm), also in the infrared range. Such lasers are particularly attractive for fiber-optic communications. Other researchers are working on semiconductor lasers with visible (red) beams.

The semiconductor laser is clearly a comer for many low-power applications, where it is a candidate to replace the helium-neon laser. What remains is to get it all together, specifically to combine a long lifetime with a good beam quality in a laser that can be simply and economically mass-produced. It looks like it's just a matter of time, at least for infrared lasers, but there are problems in trying to move to the shorter visible wavelengths.

LIQUID LASERS

If gases and solids can sustain laser action, why not liquids? Indeed, liquids can be the active media in lasers, and there's a large and important family of liquid lasers based on organic dyes. The dyes, which are actually solids at room temperature, are dissolved in a liquid (generally an organic compound, such as alcohol) to form a solution.

What makes the dye laser special is the nature of its transitions. In almost all other lasers, the laser transition is between two states at fixed energy levels, which means that the laser emission is in a fixed, very narrow, well-defined band of wavelengths. Even in carbon dioxide lasers and other types with a family of closely spaced energy levels, there are discrete transitions at distinct wavelengths. Organic dyes, in contrast, have energy levels so closely spaced that to all intents and purposes they form a continuum. This large number of levels exists because of the complexity of the dye molecules, and as a result, electronic transitions in the dye can produce a broad range of wavelengths, most of them visible.

It's possible to devise an optical system that will select a single wavelength in the laser's range. Moreover, these optics can be adjusted to tune the output wavelength continuously across the range of wavelengths possible for that particular dye. Most dyes are tunable across at best 10 percent of the visible region, but it's possible to arrange several dyes in sequence so as to be able to tune the output wavelength across the entire visible spectrum.

What makes a commercial dye laser worth its price tag of $5,000 to $50,000 is the fact that it can be tuned precisely—often with just a twist of a dial—to emit a smaller range of wavelengths than any other source, and that within this range, the light is much more intense. This can make a dye laser invaluable for experiments in such areas as chemistry and atomic physics.

Incidentally, one of the strangest lasers ever built was a dye laser. It doesn't even emit a beam, but rather a halo. Z. G. Horvath, of the Central Research Institute for Physics in Budapest, Hungary, and two colleagues at the Lebedev Physics Institute in Moscow built a disk-shaped laser that emits light along all 360 degrees of its circumference. In this device, the circumference is coated with a partially reflecting film, producing a disk-shaped laser cavity. At the heart of this weird laser is a dye that's pumped by another laser. Horvath believes it should be possible to extend his technique to make a spherical laser—one that will emit laser light from the entire surface of a ball-shaped device. But it's unclear if either a disk or a spherical laser will ever find applications outside the laboratory.

THE FREE-ELECTRON LASER

A recent addition to the laser bestiary is a strange creature known as the free-electron laser. The term *free electron* comes from the fact that the active medium emitting the light is a beam of electrons, free from atoms, which is passing through a magnetic field.

This laser requires a large electron accelerator or storage ring to provide the electron beam. Then it needs a set of large, powerful magnets to keep changing the direction of the beam. The first such laser was operated at the Stanford Linear Accelerator Center in a 1977 experiment by John M. J. Madey, a Stanford University physicist. To give you an idea of what league we're in here, the Stanford accelerator cost hundreds of millions of dollars to build, and that's just one part of the laser. You can get by with a small accelerator, however, although it's not yet clear *how* small.

What has aroused interest is the prediction that the output of free-electron lasers should be tunable across a wide range of wavelengths: perhaps from the microwave region all the way to X rays, although no *single* device would operate across that entire range.

The military is interested in this laser for an obvious reason—its high power. Originally, it was thought that a beam of electrons would itself make a good weapon. But it would be difficult to focus a beam of negatively charged electrons, which would mutually repel one another, over long distances. Transforming this high-energy, but cumbersome, beam into a laser beam would solve a lot of problems. A laser beam is easier to focus on a distant target. A series of experiments to test free-electron lasers, sponsored by the Defense Advanced Research Projects Agency, was getting underway as this book was being written.

ENTER THE X-RAY LASER (MAYBE)

There's no better way to learn how fast-moving a field is than to write a book about it. We were in the final stages of finishing this chapter when a new type of laser apparently emerged on the scene: the X-ray laser.

We use the word *apparently,* because it isn't yet clear exactly what happened. Operation of an X-ray laser at 1.4 nanometers (1.4×10^{-6} mm) was reported in *Aviation Week and Space Technology,* and it's clear that an experiment was performed by a group at the Lawrence Livermore National Laboratory in California. However, what happened is not yet certain. Livermore's official reaction is "no comment," and there's much skepticism among other researchers about some of the details in "Aviation Leak," as the magazine is called in the aerospace community. (The secrecy is due to potential military implications, which we cover in chapter 7.)

Laser researchers tend to be very cautious about reports of X-ray lasers, because this isn't the first such report. Back in 1972, a physicist at the University of Utah made a big splash by announcing that he had built an X-ray laser. He hadn't, however. The effects he thought were caused by X rays were really caused by something else. His efforts to publicize the "discovery" (which was reported in *Newsweek* and elsewhere) before thoroughly checking it out left him disgraced in the physics community.

There's another reason for skepticism too. Even in theory, an X-ray laser is very hard to build. For example, just to produce a population inversion on a 0.1-nanometer transition takes two watts per atom. Atoms excited on such a transition drop down to the ground state in about 10^{-15} sec. The very process of pumping an X-ray laser would vaporize it. And it wouldn't even be a laser oscillator, because no mirrors exist for X rays, making it impossible to build a resonator.

Why bother trying? Because it's a challenge to physicists. Of course, there are some potential applications for X-ray lasers, some of them military, a fact that physicists regularly point out to the people who fund their research. Yet the people working on the problem generally seem more interested in the intellectual challenge of finding a way of doing this very difficult task.

"Difficult" may be too mild a word. It reportedly took a small nuclear explosion to pump Livermore's X-ray laser. Much of the energy from such an explosion is in the form of X rays, and these X rays excited atoms to produce an X-ray population inversion. The whole process happened very fast, probably in about one picosecond (10^{-12} sec, or one trillionth of a second) or possibly even less, according to X-ray laser researchers outside Livermore. *Aviation Week* said that during that ultrashort pulse, the laser produced a very high level of energy—hundreds of trillions of watts. Yet because the pulse was so short, the total energy in the pulse was probably only a few hundred joules (a joule is the amount of energy provided by a power of one watt delivered for one second)—enough to keep an ordinary light bulb burning for only a few seconds.

Whether or not the X-ray laser will ever be "practical" depends on how you define that word. The need to energize it with a nuclear bomb (albeit a small one) obviously presents some serious problems. Developers of military systems have some ideas, and while their ideas sound like science fiction now, it's too early to be certain.

WHERE DO WE GO FROM HERE?

What new materials will be found to lase? We put this question to Arthur Schawlow recently. He just shook his head and said he couldn't answer the question, that scientists had proved that so many materials could be used in lasers that it was impossible to predict what would be next. He would only say that laser scientists should look for new materials and new transitions that would produce better lasers in the visible region of the spectrum. Existing visible-beam lasers are too inefficient, he said. That's a sentiment you can hear echoed in the halls of the Pentagon and the Department of Energy's headquarters, for altogether different reasons.

If Schawlow and many others get their way, the next breakthroughs in laser technology will be ones we can see with the naked eye.

4 THE SHORT BUT TEMPESTUOUS HISTORY OF THE LASER

The history of the laser has all the elements of a thriller mystery: patents potentially worth hundreds of millions of dollars, a group of brilliant rivals all vying for the honor (and in some cases the wealth) of having invented the laser, a feud between two of these scientists stretching over four decades, a gigantic legal battle that continues to this day, and a man's career put in jeopardy by military secrecy and Joseph McCarthy's red-baiting.

Add to all these ingredients the Nobel Prize for physics and an astounding revelation about the planet Mars, and you've got the improbable plot of a B-grade movie. And yet it is the real-life story of the invention and development of the laser and the principles that lie behind it.

It is basically the story of seven men. Three of them—American Charles H. Townes and Russians Nikolai G. Basov and Aleksander M. Prokhorov—shared the Nobel Prize in 1964 for their earlier work that eventually led to the invention of the laser. A fourth, Arthur L. Schawlow, collaborated with Townes to obtain the first United States patent covering the fundamental principles of the laser, and eventually shared the 1981 Nobel Prize in physics for using lasers to study the nature of atoms and molecules. Another, Theodore H. Maiman, working alone, actually built the first laser in 1960. And a sixth, Gordon Gould . . . well, it's tough to sum up Gordon Gould's role in the mystery of the laser in a single sentence. His is a complicated role, and one that gives the story much of its spice and controversy, and even a touch of bittersweetness. We will take a close look at his contribution shortly.

These men did most of their pioneering work in the 1950s. But to do justice to laser history, we have to go back sixty years, to a seventh man, one you've probably heard of.

THE MASTER STARTED IT ALL

The story begins in 1916, when Albert Einstein was studying processes involving the electrons in an atom. Normally electrons can either absorb or emit light. In fact, electrons spontaneously emit light without any outside intervention. But Einstein predicted that electrons could also be *stimulated* to emit light of a particular wavelength. The stimulus: additional light of that wavelength. Although Einstein's prediction was verified by R. Ladenberg in 1928, it was not until the early 1950s that anyone seriously considered building a practical device making use of this phenomenon.

Remember, laser stands for Light Amplification by Stimulated Emission of Radiation. Einstein discovered *stimulated emission,* but to make a laser, you also need *amplification* of the stimulated emission.

The first known proposal to amplify stimulated emission was made in a Soviet patent application in 1951 by V. A. Fabrikant and two of his students. That patent was not published until 1959, however, and hence did not influence other researchers. Fabrikant remains somewhat of a mystery to this day, one of the forgotten men of laser research. In mid-1953, Joseph Weber of the University of Maryland also proposed amplifying stimulated emission, a concept that was explored in much more detail the following year in a paper by the aforementioned Russions Basov and Prokhorov. Weber has since become better known for research in another field—the detection of gravity waves, yet another decades-old idea of Albert Einstein.

These are the official dates of the early laser race. But perhaps the most significant occurrence took place on a park bench in Washington, D.C., on the morning of April 26, 1951. Charles H. Townes was attending a meeting of physicists in Washington, sharing a hotel room with Arthur Schawlow. To be precise, Townes was attending a conference on millimeter waves while Schawlow attended another meeting. A prime interest of Townes was to generate short wavelengths for research purposes, a task he had yet to achieve. Townes, a married man with small children, was accustomed to waking up early. Schawlow, a bachelor, was not. So when Townes awoke early that morning, he went

Charles H. Townes (*left*) stands with James P. Gordon beside their second ammonia maser, which has its side removed so that the inner workings can be seen. At right rear is T. C. Wang, who is standing next to the very first ammonia maser in this mid-1950s photo. Courtesy Charles H. Townes

out for a walk, to avoid disturbing Schawlow. And so it was on a bench in Washington's Franklin Park that he had his Eureka experience. He suddenly realized what conditions would be needed to amplify by stimulated emission of microwaves. As we have seen, microwaves are very short electromagnetic waves, the same kind used in microwave ovens, for example. They are *not* light waves. Yet Townes's revelation was infinitely relevant to the laser.

The idea Townes got that day "seemed only marginally workable" at the time, in his words. In the traditional manner of physics professors, he formulated the problem as a thesis topic and turned it over to one of his graduate students at Columbia University, James P. Gordon. Three years later at Columbia, Gordon, Townes, and Herbert Zeiger had the first maser (for microwave amplification by stimulated emission of radiation) working.

In the years that followed, masers proliferated. Because the physics involved was fascinating, the field attracted many researchers. Unfortunately, there were few uses for masers. Masers make good amplifiers for boosting the signals that radio astronomers receive from outer space or for satellite communications, and they serve as frequency standards in ultraprecise atomic clocks. But they amplify too narrow a range of frequencies for most electronic applications. Physicists wanted more, and soon, they began looking to other areas of the electromagnetic spectrum, specifically to light at infrared and visible wavelengths. And thus began the great race.

RACING TOWARD THE FIRST LASER

This is where it starts to get interesting . . . and controversial. In September 1957, Townes sketched a design for his "optical maser," which would emit visible light. He then contacted his old friend Arthur Schawlow, who had by this time left Columbia to work at Bell Labs—and had also given up his bachelor status to marry Townes's sister. Together, Townes and Schawlow developed a detailed plan for building a laser.

Enter Gordon Gould. Gould was a graduate student in the physics department at Columbia, where Townes was a professor. In fact, his lab was just a few doors down the hall from Townes's office. Gould has often been described as Townes's graduate student, but this is stretching the point. Townes says that he taught a course taken by Gould, but that Gould was not "his graduate student" in the sense that he directed Gould's research. Since Gould and Townes were eventually to become involved in a patent dispute, this is a fine, but important, point. Gould was in fact the graduate student of Nobel laureate Polykarp Kusch.

Gould admits that he was inspired by Townes and his development of the maser. He was obsessed with the idea of developing a device that would emit light rather than microwaves but was unable to convince Kusch to approve the project as a Ph.D. thesis. So he decided to go it alone. And in November of 1957, just two months after Townes first sketched his design for an optical maser, Gould sat down with a notebook and wrote down his own idea of how to build such a device. Gould began his notes by coining—apparently for the first time—the name *laser*. He went on to outline his plans for construction and made some prophetic statements along the way. Gould says that he recognized, before other laser pioneers, that lasers could produce power densities previously unattainable. He noted that the second law of thermodynamics does not limit the brightness of lasers. This law holds that the temperature of a surface heated by a beam from a thermal source of radiation cannot exceed the temperature of the source. Gould realized that a laser would be a *non*thermal source of light and would thus be able to generate temperatures far greater than its own. In practice, this means that a laser operating at room temperature is capable of producing a beam that could melt steel. A focused laser beam could be used to trigger thermonuclear fusion—a process Gould predicted in his notes, where he also said that lasers could be used for communicating with the moon.

After completing his notes, Gould went to a Bronx (New York) candy store, where he had them notarized by the owner, Jack Gould (no relation). A reproduction of the first notarized page of Gordon Gould's notebook is exhibited today in the Smithsonian Institution.

At about this time, Townes telephoned Gould to ask him about the thallium lamp he was working on for his Ph.D. thesis. The excitation of thallium is related to the excitation of electrons that takes place in a laser—or, at the time, in the *proposed* laser. The timing of that conversation is important, and it's a point on which Townes and Gould disagree. Gould says that it was after he completed his notes, but Townes says his records indicate that it was about three weeks *before* Gould's first notes on lasers. Gould says the conversation warned him that Townes might be working on the same project. Townes says that he explained what he was doing to Gould, but that Gould said nothing at the time about any ideas he had.

In any case, Gould promptly took his notebook to a patent attorney, who did not understand the significance of the laser and gave Gould the mistaken impression that he would have to reduce his ideas to a more practical level to get a patent. So he didn't apply for a patent at that time (he waited until April 1959).

But Townes and Schawlow soon did. About seven months later, in the summer of 1958, they applied for patents and submitted a detailed paper to the prestigious journal *Physical Review,* which published it in December 1958. Gould, besides failing to file for a patent immediately, compounded the error by not publishing his laser proposals in a scientific journal, the standard procedure that scientists use to get their colleagues to recognize them as the originators of ideas.

Gould left Columbia without receiving his doctorate and took his laser concept along with him to TRG, Inc., a small company in Syosset, New York. TRG used Gould's ideas in a proposal it made to the Department of Defense's Advanced Research Projects Agency (ARPA), later renamed the Defense Advanced Research Projects Agency (DARPA). It was the laser's heating effects that the military was interested in, and the Pentagon was so excited by Gould's notions on how to adapt the laser for warfare that it gave TRG $1 million in 1959, rather than the $300,000 the company had asked for.

SECRECY AND RED HUNTS

The Pentagon was also sufficiently impressed with Gould's ideas to clamp a tight security lid on TRG's research. The military quickly identified one important security risk working on the secret project—Gordon Gould himself. In the early 1940s, Gould had had a brief flirtation with Marxism. Or, as he puts it, at the time he was "married to a woman who became a Communist." He is no longer married to her. This all happened while he was working on the Manhattan Project—the project that developed the atomic bomb. He and his former wife were in a Marxist study group that was run by an FBI informer. This group's leader, according to Gould, was a paid provocateur whom the FBI blackmailed into becoming an informer and getting people to join the group. Gould's interest in socialism ended in disillusionment in 1948 when the Soviet Union took over Czechoslovakia. His wife did not share his disillusionment, and they separated.

But Gould's brief association with Marxism would haunt him. He was fired from his teaching job at New York's City College in 1954, and he claims he suffered harassment from "people like [Wisconsin Senator Joseph] McCarthy." And in the Cold War year of 1959 when the ARPA contract to TRG was issued, such a history was enough to deny Gould the required "secret" clearance. He couldn't work on his own project.

Gould stayed with TRG, but he had to work physically apart from those with security clearance. There were two buildings—one for those with "secret" clearance, one for those without. Gould worked in the latter. Other researchers could ask Gould questions, but they couldn't tell him what they were doing. Gould says, though, that he got a pretty good idea of what was going on in the other building from the kinds of questions asked him. "But," he adds, "this certainly slowed things down." Gould also complains that it is difficult to attract first-rate scientists to your company when you can't tell them what you're working on.

Townes and Schawlow had no such "help" from the government and thus were able to work in peace on laser development, at Columbia and Bell Labs, respectively. Several other groups were also racing to build a laser. Remember, despite the fact that Townes, Schawlow, and Gould had all applied for patents and made various detailed proposals, and despite the fact that various Russians had done likewise, by the late 1950s no one had actually built a laser yet. At the time, gases were thought to be the best candidates for laser action. But everyone was in for a bit of a shock.

A SURPRISE IN MALIBU

Among the people watching this flurry of activity was Theodore H. Maiman, a physicist at the Hughes Aircraft Company's Research Laboratories in Malibu, California. Maiman had been working with synthetic ruby as a maser crystal and had studied the material carefully. Other researchers had largely dismissed ruby as a laser candidate because of characteristics of the atoms within the crystal, but Maiman's calculations convinced him that it would work.

Working alone and without a government contract, Maiman put together a small device in which a cylindrical crystal of ruby about 1 cm (0.4 in) in diameter was surrounded by a helical flashlamp. The ends of the ruby rod were coated to serve as mirrors—a necessity for laser oscillation. When intense flashes of light lasting only a few millionths of a second illuminated the crystal, it "lased"—that is, it produced short pulses of laser light.

On July 7, 1960, Maiman announced to the press that he had operated the first laser. It was a small thing, only a few inches long—so small, in fact, that a Hughes public-relations officer refused to photograph it for the press release. Instead, he insisted on taking a picture of a larger device that had not yet lased, but which he considered more impressive because of its larger size. That attitude may seem bizarre in this era of microcomputers and integrated circuits, but in 1960 most electronic equipment was still made out of bulky vacuum tubes, and bigger was somehow better.

Maiman's laser produced about 10,000 watts of light but only for a few millionths of a second at a time. The light was so far to the red end of the spectrum that it was barely visible. It required sophisticated instruments to show that the pulses of light were not ordinary fluorescence but were a type of light that had never been seen before, laser light. It was the start of the laser era.

Unfortunately, the implications of Maiman's discovery weren't obvious at the time to the editors of one of the most prestigious journals in his field, *Physical Review Letters.* Having decided in 1959 that advances in maser physics no longer deserved rapid publication (the major purpose of *Physical Review Letters*), the editors rejected Maiman's paper.

His second choice was the prestigious, but less specialized, British journal *Nature.* That journal promptly published Maiman's little article in 1960; at only about 300 words, it is one of the most succinct reports ever made of a scientific breakthrough. That little paper was enough to allow several other laboratories to duplicate Maiman's feat.

THE BOOM BEGINS

After studying Maiman's work, researchers' attention quickly turned to building new types of lasers. At first it was slow going. The first gas laser and two more crystalline lasers were built in 1960, one of the latter by Schawlow. Two new lasers were discovered in 1961, one of them by Gould's group at TRG, Inc. Like Maiman's laser, this was optically pumped, but its active material was the vapor of cesium (a metal).

The real boom started in 1962, and by 1965 laser activity had been observed at a thousand different wavelengths in gases alone. People began to search for ways to use lasers as soon as they had been discovered. One of the first applications was in finding how far away distant objects were, an application quickly picked up by the military for use by soldiers who need to know how far they are from targets. Researchers at Bell Labs and elsewhere began studying communications, the application first envisioned by Townes and Schawlow.

Commercial laser manufacture was not far behind. One of the first companies in the business was Korad, Inc., founded by Maiman in Santa Monica, California, in 1962. Others soon followed. Many failed, and some remain tiny enterprises with only a handful of employees. Others have been very successful, the largest being Spectra-Physics, Inc., based in Mountain View, California, which has sales of over $100 million per year and has its stock traded on the New York Stock Exchange.

Honors soon began pouring down on the laser pioneers. In 1964, Townes, Basov, and Prokhorov shared the Nobel Prize for physics. Townes received a patent on the maser which—because it covered amplification by stimulated emission regardless of wavelength—also applied to the laser. Townes and Schawlow also shared a basic patent on the laser (i.e., on a device specifically operating at optical and infrared wavelengths). Maiman was granted a patent on the ruby laser and eventually realized a tidy sum by selling his interest in Korad, Inc., to the Union Carbide Corporation.

THE RETURN OF GORDON GOULD

Gordon Gould, meanwhile, seemed to fade away. Townes and Schawlow held the patent he wanted, beating him to the punch with their application by about nine months. Trying to make his own 1959 patent application stick, he got involved in five costly and time-consuming interference actions, which are procedures used by the U.S. Patent Office for determining which party is entitled to a patent on a particular invention. The first interference action pitted Gould against the Townes-Schawlow patent. He essentially lost the decision and also succeeded in creating a stir in the scientific community because of the stature of the men he was taking on. Gould went on to lose two more such battles, but he won two others, which many years later would become the basis for patents issued to him. By the time it was all over, his company had dropped a total of $300,000 on lawyers. And most of his claims had been forgotten. In 1970, he got his patent rights back from the company and began pushing the applications himself. Eventually, unable to finance any more battles in the patent war, Gould signed over part of his interest in his patent position to the Refac Technology Development Corporation, a New York patent-licensing company, in return for Refac's agreement to press for patent licenses.

Refac got results. On October 11, 1977, Gould was issued a patent on optical pumping, a technique needed to make many lasers work, as explained in chapter 3. In 1979, Gould received a second patent, which, like the optical-pumping patent, was an outgrowth of his 1959 application; this second patent covers a broad range of laser applications.

The laser industry was shocked when Gould received his optical-pumping patent. The Townes-Schawlow patents had just expired, and laser manufacturers thought they were through with paying royalties on basic laser concepts. Affected by the ruling were manufacturers of the many industrial and low-power military lasers that rely on optical pumping. When Refac demanded royalties of about 5 percent, laser makers vowed to fight the patents. The first suit charging infringement of the optical-pumping patent was brought against the Control Laser Corporation of Orlando, Florida, before the ink was dry on the patent, but nearly four years later it had yet to go to trial.

The case is a good bet to end up before the Supreme Court. It's one of the most complex patent cases in the history of the United States. The 18 pages of fine print in the patent disclosure are accompanied by a 500-page "file wrapper" detailing the patent's legal history, which must be studied thoroughly to determine the patent's validity.

The complexities of the case include both sticky technical questions and fine points of the law. To be valid, a patent must contain sufficient detail to enable anyone with sufficient skill and resources at the time the application was filed to build the device described in the patent. As Maiman has pointed out, neither Schawlow, Townes, nor Gould had built a laser when they applied for their patents, or even shortly thereafter. However, more than twenty years after the fact, in early 1981, Gould and a colleague built a laser relying only—according to Gould—on information in his patent application and on other information and equipment generally available when the application was filed in 1959.

No sooner had Gould built his laser and offered to bring it into the courtroom than another complication emerged. Detection of optically pumped laser amplification in the atmosphere of Mars was reported in the April 3, 1981 issue of *Science* by a group led by Michael Mumma of NASA's Goddard Space Flight Center. Mumma's group found that sunlight produces a population inversion in carbon dioxide some 75 to 90 km (45 to 55 miles) above the surface of Mars, producing amplified stimulated emission (what we call laser amplification) in the infrared. Control Laser Corporation greeted the discovery with glee, saying that it demonstrated that optically pumped laser amplification was a natural phenomenon and therefore unpatentable. There the matter rested as we put the final touches on this book, leaving the courts with a truly massive tangle of legal and technological questions to resolve.

Complex issues also surround Gould's more recent applications patent. This case began with Refac suing Lumonics Research, Inc., a small Canadian company that makes laser systems that can etch markings on objects. The General Motors Corporation then intervened on behalf of Lumonics and now seems to be carrying most of the defense load. GM argues that the patent is invalid, because it's an obvious extension of "prior art," extending back to "the use by Archimedes in 212 B.C. of a burning glass to set fire to Roman ships in the siege of Syracuse." That's "patently ridiculous," counters Eugene M. Lang, president of Refac.

There are a couple of final ironies here: Townes is a director of General Motors, although the company insists that he was not involved in GM's decision to intervene in the suit. And it was Townes and two coworkers who in 1973 first observed the infrared emission from the atmosphere of Mars that further observations in 1980 by Mumma's group showed was due to laser amplification.

SENSITIVE ISSUES

One factor that helped delay recognition of Gould's contribution to laser development was his failure to follow the standard procedures of the scientific community. Although scientists are expected to be concerned with patent rights, they are also expected to describe their research promptly in a scientific publication, in order to both inform other scientists and establish the priority of their work. In explaining his failure to follow standard publishing practices, Gould cites time pressures, the potential conflict of publishing with filing foreign patent applications, and the fact that, thanks to the military, much of his information was classified. This failure has clouded his role in the history of the laser and continues to do so.

There's a more sensitive issue involved, too: the treatment of graduate students. Many graduate students get their research ideas from their professors, but a few faculty members are not above appropriating their students' ideas as their own. Townes claims that most of the ideas in Gould's notebook and patent application were outgrowths of Townes's earlier description of his concepts to Gould. Gould claims his ideas were original. The courts may have an opinion on the matter, but it's unlikely that it will ever be definitively resolved.

On a personal level, there is still obvious hard feeling between the two men. Townes told us recently that he feels that there are many people who have contributed greatly to the development of the laser. But he went on to say that Gould wasn't one of the important ones. Gould describes Schawlow as "a very nice guy" but refuses to characterize Townes, making only the cryptic comment, "He has his needs, I guess." When we asked Schawlow what he felt about Gould, the normally jovial physicist became visibly shaken and admitted he was bitter over Gould's patent claims.

Issuance of the patents has given Gould both emotional and financial satisfaction. He finally realized a profit on his investment in the laser patents by selling off part of his remaining interest for $300,000 in cash and notes worth $2 million. The buyer was another strange player in the patent game, the Panelrama Corporation of Ardmore, Pennsylvania, which liquidated its money-losing chain of retail building-supply stores to make the purchase. Panelrama then changed its name to the Patlex Corporation, because its interest in the Gould patents is essentially its only business. If the patents are upheld, Patlex, along with Gould, Refac, and the New Jersey lawyers who handled the case in return for an interest in the patents, will share royalties that could run into tens or even hundreds of millions of dollars over the life of the patent. Gould himself estimates that the patents could be worth $10 million a year. And a patent is good for seventeen years. Many observers in the laser community, however, believe that the applications patent is too vague to be valid, and that even the optical-pumping patent may not withstand a concerted legal challenge.

Like most of the other laser pioneers, Gould has gone on to other things. Now over sixty years old, he's vice-president of Optelecom, Inc., a small company in Gaithersburg, Maryland, that manufactures equipment for fiber-optic communications. Optelecom's largest customer is the military, but Gould is hoping to see the day when his patent income will let him pick his own research, without regard to what the military wants. Now that his patents have been issued, Gould has begun receiving awards, such as the Inventor of the Year award presented by the Association for the Advancement of Invention and Innovation. But it's hardly a matter of pride with him any more. "I don't have any ego problem with respect to these patents," he says, "but I would like to see some money out of them."

Townes and Schawlow have enjoyed successful careers in academia, each receiving awards too numerous to list here. Townes, now professor of physics at the University of California at Berkeley, has for some time been occupied primarily with radio and infrared astronomy, using masers and lasers for some of this work. Schawlow, now a professor of physics at Stanford University, uses lasers as tools for studying the properties of matter instead of studying lasers themselves. For that work Schawlow and Nicolaas Bloembergen, a Harvard University physicist who was also active in the early years of laser development, shared the 1981 Nobel Prize for physics. At last, Schawlow joked when we spoke to him the morning that the award was announced, he would no longer have to explain that he hadn't received the Nobel Prize—a common misperception because of his close involvement with Townes in developing the laser.

After his years at Korad, Inc., Maiman, too, drifted away from lasers. He tried commercial ventures in other fields and worked for several years as an independent consultant before joining TRW, Inc. as vice-president for technology and new ventures. Many other early figures in the field, such as Gordon, Zeiger, and Weber, also are no longer active in laser research.

Of all the prominent laser pioneers, Basov and Prokhorov have remained the closest to laser research. Basov is director of the Lebedev Physics Institute in Moscow, where much laser research is performed, and is also a member of the Soviet Parliament. Prokhorov is deputy director of Lebedev. Both men head large groups engaged in laser-related research, and both their names appear regularly on reports of that research.

5 LASER MEDICINE: A BRIGHT PROMISE

It's an old, and not very funny, joke among medical writers. You always get two big stories out of every new medical development. At first, you get to write an optimistic story: "The Promise of ______." You can fill in the blank with any new drug or procedure, whether it be the swine flu vaccine or radical breast surgery. Then, a few years later, you must always write another story: "The Dangers of ______." And you fill in the blank with the same words.

We don't want to do that here. We don't want to fill you with unrealistic hopes that will be dashed ruthlessly a few years hence. Lasers are, in fact, a new hope for many patients. But that bright promise is still just that, a promise. The laser has proved valuable to doctors in some areas—most notably in treating certain eye conditions and for gynecological surgery. But it is hardly the cure-all many recent articles would have us believe.

Twenty years after the invention of the laser, the medical profession is still trying to decide what to do with it. Lasers have been tried in an amazing variety of medical applications, from surgery to bleaching out tattoos, from acupuncture to melting material for dental fillings. There is a recent report from Israel that doctors there have used a laser to circumcise a 12-year-old hemophiliac.

The problem is that lasers must compete with a vast array of other techniques that the medical profession has developed. Unless lasers offer a significant advantage over other techniques, they're not going to be used. One reason is simple inertia: physicians are comfortable with existing methods. A more substantial reason is cost. A laser for general surgery costs $50,000 or more, a thousand times the price of a good scalpel. A laser system for treating eye conditions costs about $30,000.

One of the cheapest medical laser systems, at $3,500, is an acupuncture system. But that's still expensive compared to needles.

The laser "will not replace the scalpel entirely, nor will it supplant electrocautery [use of an electrical discharge to burn away tissue and seal blood vessels]," wrote Richard M. Dwyer, a member of the clinical faculty at the medical school of the University of California at Los Angeles, in a review of laser applications in medicine. "It will, however, provide a treatment procedure for certain problems which will result in improved patient care," he continued. The facts bear out Dwyer's assessment.

We'll describe some fascinating and important research in the pages that follow. Because it deals with health care, this chapter will touch you more personally than anything else in this book. You should remember, though, that when we say that something *can* be done with a laser, it doesn't mean that you should rush out and ask your physician to do it to you. Much of what we'll describe is experimental and needs further testing before it can be put into common use. Furthermore, using lasers for medical purposes requires special training that most doctors lack; even the most gifted surgeon couldn't operate on you successfully with a laser without such training. If you read about a procedure here and contemplate getting the treatment, we suggest you first ask your regular doctor where in your area you can find a specialist in laser medicine. If he doesn't know, a good bet is to contact the nearest teaching hospital (a hospital connected with a medical school) for help. Often these are the places where experimental and clinical work with lasers is being carried on.

THE BIRTH OF LASER SURGERY

One of the first centers for medical laser research was founded at the University of Cincinnati College of Medicine in the early 1960s by Leon Goldman, a dermatologist who soon became a vocal advocate of laser medicine. Well into his 70s, Goldman now works at the Jewish Hospital of Cincinnati, where he continues to operate with, and promote the use of, lasers. He is president of the American Society of Laser Medicine and Surgery, a group he is largely responsible for organizing.

Early experiments were crude (so was the hardware, by today's standards). To see what a laser would do to human skin, Dr. Goldman directed a laser beam at his arm. He did this hundreds of times and performed other experiments on his patients, thus helping identify many key lines of research that are still being pursued today.

Laser surgery was one of the first ideas that occurred to physicians, particularly for delicate operations that couldn't be performed with a knife. Goldman and others soon found that they could clear up many skin disorders with laser light. Other possibilities opened up, such as laser-assisted healing of wounds or ulcers and laser techniques for diagnosis and measurement.

Today, Goldman bristles at the mention of the idea that laser medicine has yet to prove itself: "For operations on the larynx, gastrointestinal surgery with bleeding [such as a bleeding ulcer], inoperable birthmarks, and certain brain tumors, lasers are *required.*" And when Goldman says "required," he means that lasers not only *can* be used for these procedures, but that they *must* be used. Goldman admits, however, that he is biased, and that these are recommendations from his own organization, the American Society of Laser Medicine and Surgery. He also says that the laser is "obligatory" in gynecological surgery, specifically in treating early invasive cancer in women. Even Goldman, though, has a caveat for all patients and physicians: "If you don't need the laser, don't use it."

Surgery is the most dramatic medical use of lasers and was shown in the motion picture *Logan's Run.* But the reality wasn't quite dramatic enough for Hollywood. One of the major attractions of laser surgery is that the laser seals off small blood vessels as it cuts, thereby preventing most bleeding. So why did patients in the movie bleed during surgery? Because the moviemakers, despite the protests of the film's laser consultant, Chris Outwater, decided that where there's surgery, there's blood.

In real laser surgery, there is almost no blood because of the way the laser beam interacts with tissue. This is why the Israeli doctors used a laser to circumcise that boy—because the boy was a hemophiliac, his blood would not clot properly if he were cut with a knife. Most laser surgery is performed with carbon dioxide lasers emitting around 50 watts of light at a wavelength of 10 micrometers in the infrared. This is not much more power than an ordinary incandescent bulb emits in the infrared and visible regions combined, but the entire beam can be focused on a spot as small as about 40 micrometers (0.04 mm). Furthermore, light at a wavelength of 10 micrometers is strongly absorbed by the water in living cells. The combination of high intensity and high absorption is strong enough to actually vaporize cells in the focal spot.

There are two ways a surgeon can use a laser. To cut deeply, the surgeon focuses the beam to a fine point, which he moves slowly along a line, so that it cuts by vaporizing a thin line of tissue. To burn away surface tissue, the surgeon focuses the laser less intensely and scans the beam more quickly back and forth over the tissue to be removed. Fortunately, human cells are poor conductors of heat, so the tissue surrounding the area irradiated by the laser beam remains undamaged. The vaporization process automatically seals off (or cauterizes) most blood vessels smaller than about 0.5 mm (0.02 in) in diameter, thus preventing bleeding. Larger blood vessels must be closed off, however, as in conventional surgery.

The laser itself is too bulky for the surgeon to pick up and therefore remains stationary. A special arm—similar in appearance to the arm that connects a dentist's drill to the chair—transmits the beam from the laser to a beam-focusing tool that the surgeon holds in his hand, much as he would hold a scalpel. The surgeon can turn the laser on and off with a foot switch.

The arm used with a carbon dioxide laser is a jointed assembly of rigid tubes containing a series of mirrors that direct light from one mirror to the next and eventually to the patient. A simpler arrangement can be used with lasers emitting at much shorter wavelengths, such as argon lasers, which emit a blue-green beam, and neodymium lasers, which emit invisible infrared light. With these, a flexible tube containing a bundle of special glass fibers can be used, instead of the awkward, multielbowed arm. The beam is delivered to the operating site through a focusing head at the end of the tube. Although such fiber bundles are more convenient than multijointed arms, at present they can't transmit light at the wavelength of the carbon dioxide laser.

There's another problem with the carbon dioxide laser. The beam is invisible, and the surgeon needs to know where it's going. So makers of surgical lasers use a second laser, a small helium-neon model that emits a low-power red beam. The two laser beams pass along the same optical path from mirror to mirror and are focused on the same spot. The low-power red beam tells the surgeon where the invisible infrared beam will cut when it's turned on.

Most surgical lasers come with other vital accessories. A variety of

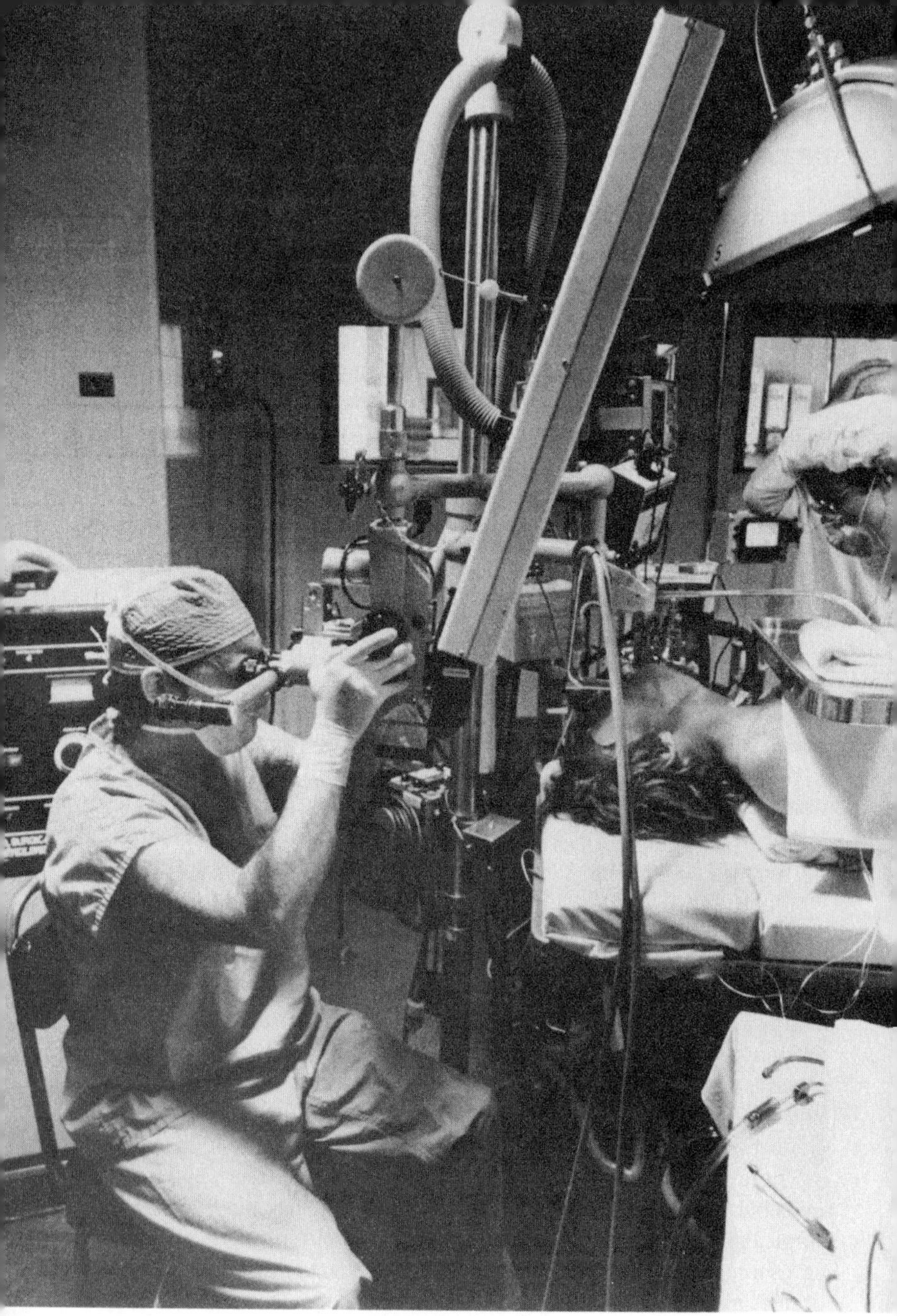

A doctor performs surgery with a large carbon dioxide laser. Since the beam from such a laser is invisible, a red helium-neon laser beam is used to pinpoint the target area. Courtesy Coherent, Inc.

focusing heads may be needed, as each head is often designed for a single type of surgery. These heads are light and are as easy for the surgeon to use as a scalpel. Microscope attachments let the surgeon take advantage of the fine focusing possible with a laser to perform microsurgery. A third accessory is a vacuum attachment to suck up steam, smoke, and other airborne debris produced when the laser vaporizes tissue.

All this hardware is expensive. While a plain 50-watt carbon dioxide laser costs only $10,000 to $20,000, a whole system built around it will run $50,000 and up because of the accessories and special features—what engineers call the "bells and whistles." Although these price tags keep such equipment out of most surgeons' offices, surgical lasers are not rare in teaching hospitals associated with major medical schools.

The high cost of the equipment tends to limit its use to types of surgery where the laser has major advantages over conventional procedures. These include microsurgery and operations on tissues rich in blood vessels and on areas where there isn't enough room to use an ordinary scalpel but where there is room to use a laser. Prominent examples include gynecological surgery and operations on the mouth, nose, and throat. The operations themselves are like any other surgical operations, except that they tend to be performed on a small scale only—in the throat or ear, for example, as opposed, say, to open-heart surgery. The patient may be given either a general or a local anesthetic, as in a conventional operation. Some laser operations, notably eye and gynecological operations, are often done in a doctor's office. If you were to witness a laser operation, you would find one striking difference from ordinary surgery. As mentioned before, there would be little bleeding; instead, you might see smoke rising from vaporized tissue. You might also see a bit of scattered light if a visible laser beam was being used as either the "cutting" tool itself or a guidance beam.

GYNECOLOGICAL SURGERY

The tissues of the cervix and vagina are particularly vulnerable to cancer and other diseases. Unfortunately, they're also hard for a surgeon to reach, very sensitive, and rich in blood vessels. To make things worse, some cancers and precancerous conditions spread out over the surface rather than localizing themselves in a single tumor.

Just the job for a laser. In conventional surgery, a scalpel must cut underneath the tissue to be removed, creating a large wound and often heavy bleeding. With a laser, the surgeon merely scans the beam across the affected area slowly enough to burn away the diseased tissue. The laser beam can reach places the scalpel can't, so the patient doesn't have to be opened up as much for good access. The laser also cauterizes as it cuts, sealing blood vessels to prevent much bleeding and deadening nerves to minimize pain (of course, the patient is anesthetized during the operation, but it's still best to minimize trauma during the procedure, so as to cut down pain during the recovery period).

The rare form of cancer often found in women whose mothers took diethylstilbestrol (DES) during pregnancy to prevent miscarriages can be treated effectively with a laser. During their teens and twenties, many women who were exposed to DES before birth (often referred to as DES daughters) develop abnormal cells, often considered precancerous, in their vaginas; in many cases this condition is followed by vaginal or cervical cancer. Laser irradiation is now an accepted method for getting rid of these abnormal cells and cancers, which are generally located at, or near, the surface of the tissue.

"Most [laser] gynecologic surgery can be performed on an outpatient basis," without hospitalization, according to Michael S. Baggish, a specialist in obstetrics and gynecology at Mt. Sinai Hospital in Hartford, Connecticut. In a paper summing up three years of laser treatment of some 300 patients, he wrote that recurring disorders, such as those related to DES, can be treated repeatedly with a laser, "with virtually no scar formation." He concluded that "the excellent functional and cosmetic results offer emotional benefits, as well as physical advantages, to the laser-treated patient." He also said that because laser surgery is simple and requires only local anesthesia in these cases, it can be used to treat patients who are pregnant or too ill to withstand general anesthesia or extensive surgery.

THROAT AND EAR OPERATIONS

Like the vagina, the throat and ear are difficult places for the surgeon to use his scalpel. Organs in these regions are also delicate and easily injured by conventional surgery.

A particularly small and delicate organ is the larynx, or voice box.

Damage to the tiny structures within it can severely impair speech. Yet physicians treating cancer of the larynx often had no choice but to remove the entire organ. Now lasers have become accepted for surgery on the voice box.

With a laser, a surgeon can vaporize lesions, including cancers, as small as 1 mm (0.04 in) across, without damaging the rest of the larynx. This lets patients speak in a manner that is more intelligible than the coarse whisper that is all that remains after the larynx is removed entirely.

Lasers can also be used inside an even smaller body cavity, the ear. Rodney Perkins, a Palo Alto, California, surgeon, says that with one type of laser microsurgery, he's restored the hearing of about twenty patients. And he's working on other surgical techniques.

Perkins uses an argon laser, which emits a blue-green beam, rather than the infrared carbon dioxide laser that is used for the commoner types of surgery described previously. The carbon dioxide laser is more powerful, but the argon laser is adequate for delicate ear surgery. It's also easier to use. Perkins cites as advantages the better maneuverability of the smaller optics that can be used at visible wavelengths, the smaller focal spot, better blood coagulation, and the greater accuracy in aiming a visible beam. It's also cheaper. While an argon laser costs about as much as a plain carbon dioxide laser, the optical accessories are much simpler and less expensive, holding the overall cost of the system to the $30,000 range, compared to the carbon dioxide system's $50,000.

Perkins restores hearing by vaporizing defective stapes bones in the inner ears of his patients. The stapes bone is one of a series of bones that transmits vibrations to the inner ear. Hardening of the bones in the ear can cause the stapes to lock in place, severely impairing hearing.

Previously, surgeons had actually used tiny picks and chisels to remove the locked bones, but this procedure could itself damage hearing permanently and cause prolonged dizziness. Now, Perkins vaporizes part of the bone with the laser, then gently removes the rest with tweezers. He says that besides avoiding damage to the delicate inner ear, the laser method reduces bleeding, both because the beam tends to cauterize the wound, and because it requires a smaller surgical opening than conventional methods. Thousands of such operations are per-

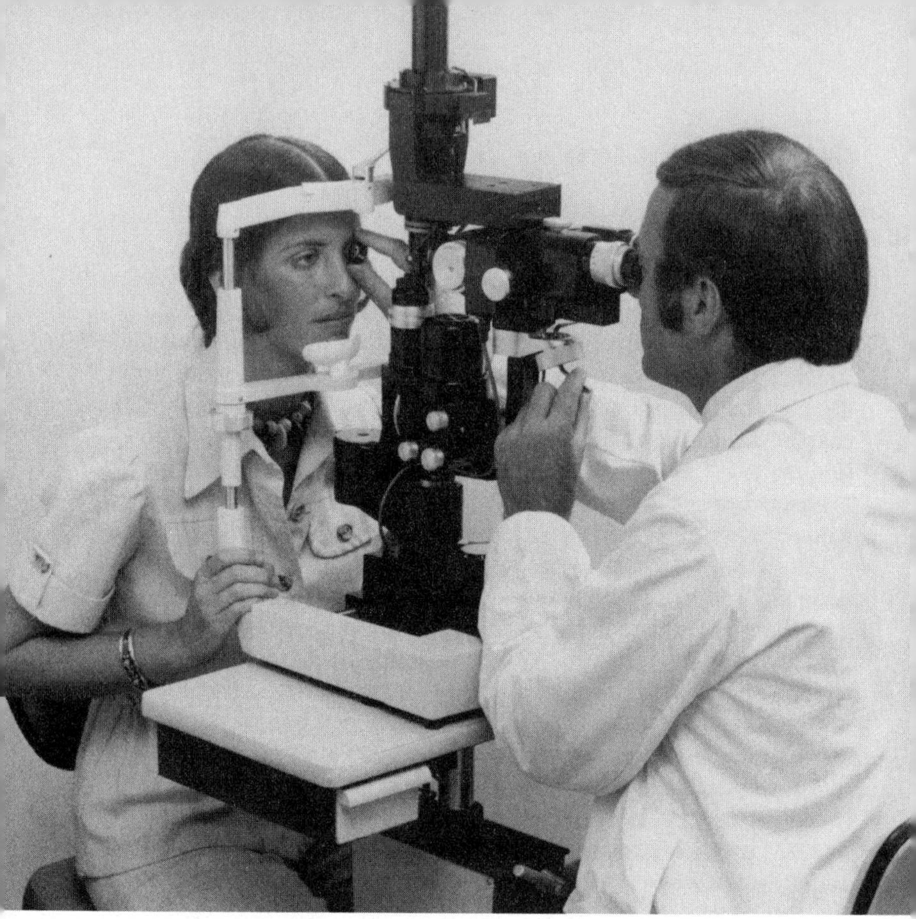

This apparatus contains an argon laser for treating patients with diabetic retinopathy, the leading cause of blindness in the U.S. Courtesy Coherent, Inc.

formed in the United States each year, but this recently developed technique cannot yet be called widespread.

PREVENTING BLINDNESS: TWO EYE OPERATIONS

The laser has proved to be an effective tool in at least stalling the progress of a disease that has become the commonest cause of blindness in the United States. The disease is diabetic retinopathy, so named because it is a degeneration of the retina, the light-sensitive area at the back of the eyeball, which affects people suffering from diabetes. The loss of vision comes from a proliferation of blood vessels on the surface of the retina. The blood vessels can obscure vision either by covering the retina so that light cannot reach it or, because they are thin-walled and fragile, by rupturing and spilling blood into the fluid inside the eyeball, thereby possibly obscuring all vision.

The solution is to destroy these excess blood vessels by directing a laser beam, with a carefully controlled exposure, at the affected area of the retina. Specialists are still debating how the laser technique works. One theory is that the beam is strongly absorbed by the blood, and the light causes coagulation in the blood vessels, thereby destroying them. Generally, an argon laser is used, because its blue-green light is strongly absorbed by blood vessels.

The treatment is far from 100 percent effective. We know two diabetics who became functionally blind by age 30 despite laser treatment. However, a nationwide study by the National Eye Institute, involving over 1,700 patients at 15 medical centers, found that over a 2-year period, laser treatment could reduce the incidence of blindness in those suffering diabetic retinopathy by a little more than half. There's nothing miraculous about the laser; the same study found that an intense flash from a xenon arc lamp was about as effective as a laser beam.

Nonetheless, the laser is preferred by ophthalmologists, probably because it is easier to control. The instrument costs about $30,000 and can be readily used for outpatient treatment in a doctor's office.

A similar technique is used to treat a detached retina, a condition in which the retina has broken loose from the back of the eyeball. Before lasers, this was a hard problem to treat, requiring either immobilization of the eye or delicate surgery. Some doctors tried to produce pulses of light intense enough to cause photocoagulation, thus creating a scar that would weld the torn retina to the back of the eyeball. Shortly after the laser was developed, W. M. Zaret found that it could do this job much more easily.

As in the treatment of diabetic retinopathy, an argon laser is used.

ZAPPING BLEEDING ULCERS

Patients suffering from severe intestinal bleeding can now benefit from two light-related inventions: the *endoscope* and the laser. The endoscope is a light pipe that allows the doctor to see inside a patient's stomach and other parts of the body. Although the idea of the endoscope is not a new one, early models were too rigid to be practical. They did not become useful until the development of *optical fibers*—fibers made of glass, designed to conduct light along their entire length even when bent. (You'll read a lot more about optical fibers in chapter 6.) Fiber-optic endoscopes are small and flexible enough that a doctor can insert one down a person's throat, to look at the inside of his stomach, or up through his anus, to examine the intestines. A regular light source is shone down some of the fibers from the outside, to illuminate the inner regions of the body. The doctor looks at the light reflected from the inside of the stomach (or other organ) through other fibers in the bundle.

When treating ulcers, doctors first used endoscopes just to find out the nature and extent of the problem in the patient's stomach. But the condition itself had to be corrected by conventional surgery.

The problem is that the prognosis for conventional surgery is often not promising. Patients suffering from severe intestinal bleeding are often in no condition to survive major surgery. Most nonsurgical techniques are ineffective.

A new solution is to use the endoscope not only as a diagnostic tool, but as a surgical tool as well, since the fibers will transmit light from some types of lasers as well as ordinary light. The doctor lines up the problem ulcer through the endoscope using ordinary light, then shines a laser beam through the fiber optics to coagulate the blood and stop the bleeding.

The optical fibers currently available work best with light in and near the visible range of the spectrum, which means that the standard surgical carbon dioxide laser cannot be used. Internal laser photocoagulation was developed independently at the University of California at Los Angeles Medical School by Richard M. Dwyer and colleagues, and at the Innenstadt Medical Clinic in Munich, West Germany, by Peter Kiefhafer, G. Nath, and K. Moritz. The first experiments were with an argon laser, but the West Germans soon found that they could get better results with a neodymium-YAG crystalline laser. The laser technique for treating intestinal bleeding is finding greater acceptance in Western Europe than it is in the United States, at least in part because of more restrictive safety regulations in the U.S.

The price of a neodymium-YAG laser system for gastrointestinal photocoagulation is slightly under $50,000. But that has to be balanced against an estimated 6,000 deaths caused by severe gastrointestinal bleeding each year in the United States alone.

REMOVING BIRTHMARKS AND TATTOOS

Laser-medicine pioneer Leon Goldman is a dermatologist, and dermatologists as a group were among the first to see the medical potential of the device.

Goldman followed the time-honored tradition of self-experimentation, as he conducted laser tests not only on his patients but also on himself. Still, some areas of laser dermatology remain controversial. One example is laser treatment of portwine stains, deep-red birthmarks on the skin, which can be disfiguring when they appear on the head or the neck. These stains are red, because they contain an abnormally dense network of blood vessels. In many cases, photocoagulation with an argon laser can make some of these blood vessels close, reducing the redness. On the other hand, laser treatment can sometimes produce undesirable coloring, even scarring, in fair-skinned people.

You can also remove tattoos with a laser. The beam breaks down or bleaches the dyes in the tattoo, which are darker and therefore absorb more light than the surrounding skin. Results are comparable to those using other techniques. However, the other treatments, which include abrading away the pigmented skin, surgical removal, and burning out the tattoo with acids, can have such severe side effects that the laser becomes an attractive alternative. Which is not to say it has no side effects. As with portwine birthmarks, the laser treatment can sometimes cause undesirable coloration or scars.

Despite Leon Goldman's claim that he has successfully treated hundreds of cases of inoperable portwine stains, the central problem in the laser treatment of both portwine stains and tattoos is one that's common to many areas of medicine: there's no way to be certain when the treatment will produce the desired result.

FAST HEALING OF WOUNDS

Illumination with a low-power laser beam can help speed the healing of some wounds on the skin. More than a hundred chronic skin ulcers—sores that do not heal by themselves—were cured by repeated

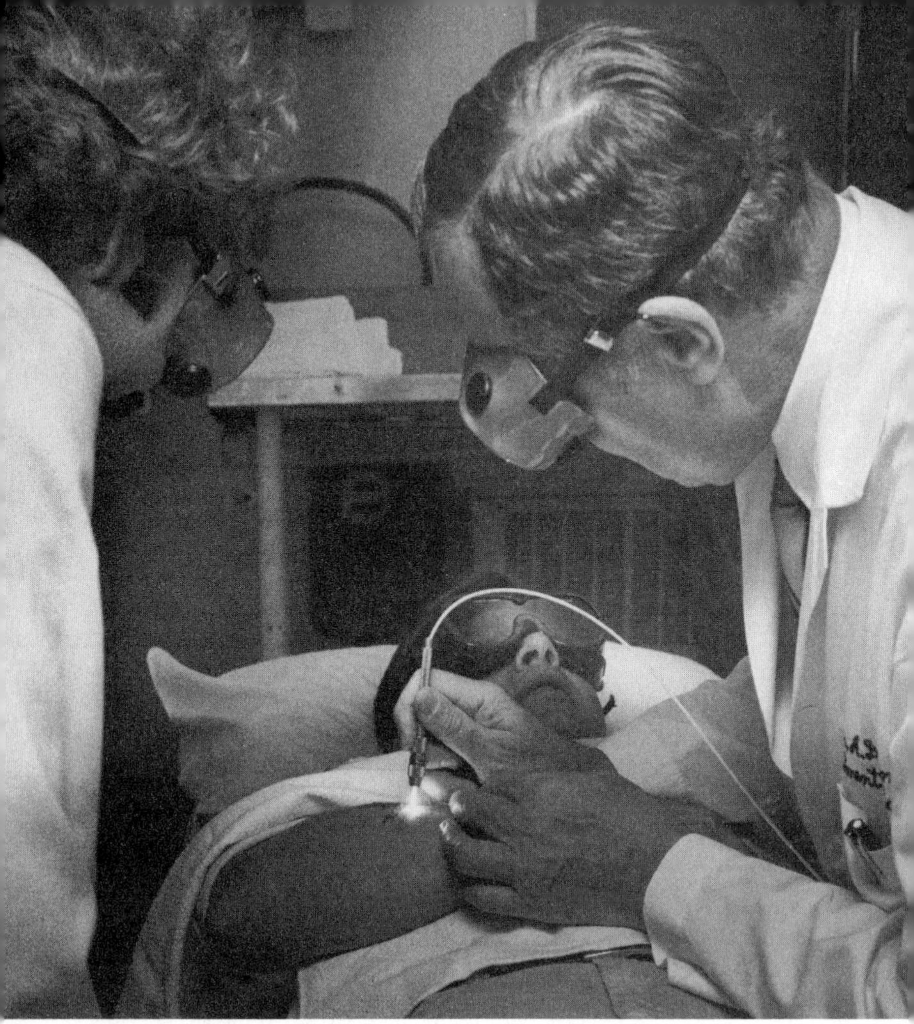

A doctor uses a laser to remove a Mickey Mouse tattoo from a patient's arm; the laser beam vaporizes the dye in the tattoo. During the procedure, special goggles are worn to prevent accidental exposure to laser light. Courtesy University of Utah Medical Center

exposure to a low-power laser at the Surgical Clinic of the Postgraduate Medical School in Budapest, Hungary. The technique was developed at the clinic by Endre Mester, who uses helium-neon lasers emitting red light or argon lasers emitting blue-green light. One problem is the need for long exposures: each treatment can last up to one hour, treatments must be repeated at least weekly, and a full series of treatments can last as long as several months.

Laser illumination can also strengthen the scar tissue over wounds on the skin. The results of Mester's experiments on rats have been supported by similar experiments performed by Kendric Smith of Stanford University's Department of Radiology. However, the mechanisms of both wound healing and scar-tissue strengthening effects are not yet understood, and the techniques have not been put into widespread use.

LASER ACUPUNCTURE

Strange as it may sound, low-power laser beams are being used to stimulate the pressure points of traditional acupuncture. The technique is finding wide use in Western Europe and, to a lesser extent, China. Conventional needle acupuncture came into vogue in the mid-1970s, after China opened up communications with the rest of the world. Laser acupuncture followed, although apparently it was not exclusively a Chinese development. An indication of the popularity of laser acupuncture is the fact that some 2,000 laser systems for performing acupuncture have been sold, mostly in Western Europe.

Laser acupuncture is performed in several places in China, although the procedure seems largely confined to the larger cities, notably Shanghai, a center of laser research, and Peking (Beijing). Some Chinese practitioners say that lasers are more effective than needles. When acupuncture is applied to pressure points on the toes, to get a fetus to rotate into the proper position for birth during the final trimester of pregnancy, for example, Chinese researchers report success rates of 80 to 85 percent with lasers, compared to 70 percent with needles.

Laser acupuncture has also helped patients suffering from chronic bronchitis and bronchial asthma. Liu Pingyi and three co-workers at the People's Hospital of Peking Medical College claim that laser treatment relieved asthmatic symptoms in 72 percent of the patients they treated. The best results were in patients under 30 years old, 90 percent of whom showed improvement. These results are preliminary; as of mid-1980, only 58 patients had been treated.

Why use a laser instead of needles? Johannes Bischko, director of the Bolzmann Institute of Acupuncture in Vienna, says that the method is popular because it is painless, hygienic, and quick. He considers laser acupuncture the only form of acupuncture suitable for children and nervous adults. The popularity of acupuncture in general in Europe is indicated by the number of patients. The Poliklinik, a central state hospital in Vienna, treats about 10,000 to 15,000 patients by acupuncture each year. About 30 percent of these are treated with lasers rather than needles, according to Dr. Bischko, who operates an outpatient acupuncture service at the hospital.

The laser powers used are generally so low that the patient doesn't even feel any warmth from the beam. No one can explain how a laser beam simulates the effects of an acupuncture needle. But then, the physiological mechanisms by which conventional acupuncture with needles can do such things as cause a fetus to rotate in its mother's womb are also unclear. Some practitioners of needle acupuncture remain skeptical of the laser approach. Many more-orthodox physicians remain skeptical about the value of *any* type of acupuncture, and laser acupuncture has not been officially recognized as an effective technique by the U.S. Food and Drug Administration. However, the FDA can regulate only sales of medical instruments, not the actions of individual doctors.

The major maker of laser-acupuncture equipment is Messerschmitt-Bölkow-Blohm of Munich, a West German company better known for its aerospace products. The system, built around a low-power helium-neon laser that costs only a couple of hundred dollars, is priced at about $3,500. The company has sold equipment to many countries, including West Germany, Switzerland, Austria, Britain, France, the United States, South Korea, and Japan. Laser-acupuncture treatments typically involve three or four sessions, with the patient billed anywhere from $10 to $100 per session.

LASER PAP SMEARS

Lasers are now being used for diagnosing disease as well as treating it. Cervical cancer is one of the commonest types of cancer in women, and early detection is critical to stopping it. The Pap smear was a breakthrough in this area: a doctor takes a tissue sample from the cervix, and this is examined under a microscope in a laboratory for early signs of cancer. But there are problems. A very large number of samples must be analyzed by skilled technicians just to detect a few cases of cancer, and it is sometimes difficult to distinguish between severe inflammation of the tissue and the early stages of cancer. You can imagine the needless distress that a false alarm can cause.

At the University of Rochester, Leon Wheeless and colleagues have a solution. They start with a traditional tissue sample from a Pap smear but stain it with a fluorescein dye. This dye, which fluoresces when struck by laser light, is absorbed more strongly by cells that have large quantities of DNA than by normal cells. Because cancer cells are constantly producing DNA and dividing, this means that they can be detected by lasers.

Wheeless's group passes a narrow stream of liquid containing some of the stained cells through a laser beam. By monitoring the fluorescence produced at various wavelengths, they can detect cells that are cancerous or have any of a number of other abnormalities.

The Rochester technique is intended for prescreening large volumes of Pap smear samples, to sort out those with abnormalities, which need examination by a skilled technician in a laboratory. Because most samples show no abnormalities, this reduces the work load of technicians considerably. Wheeless says that he has demonstrated that the new method is sensitive enough to detect cancer cells without missing any. He has now set himself the goal of reducing false alarms—normal cells incorrectly identified as abnormal—from the present level of one in about 1,200 to one in 2,000 or more.

The same basic phenomenon, fluorescence from a fluorescein dye absorbed by cancer cells, is used in a dramatic new Chinese technique. Unlike the Rochester group, however, the Chinese researchers, working in a Shanghai clinic, are using a laser directly on patients. Women are told to swallow a capsule containing the dye, and several hours later their cervixes are scanned by a beam from a low-power laser. Strong fluorescence indicates cancer. The test could be done right in a doctor's office.

The Chinese technique can readily and reliably distinguish between cancerous growth and inflammation and is simpler to interpret than the Pap smear test, according to American physicians who recently visited China. There's a potential problem, however: fluorescein dyes are mildly carcinogenic themselves. American researchers who earlier considered a similar approach discarded it because of the concern over dye carcinogenicity, according to Wheeless.

CANCER TREATMENT

Earlier we mentioned how lasers could be used as scalpels to remove certain types of cancerous tissue. Another laser approach to cancer treatment uses a method similar to laser methods of cancer detection. At the Roswell Park Memorial Institute in Buffalo, New York, Thomas Dougherty has performed experiments in which dyes are injected into animals and even humans. Once the dye has been preferentially absorbed by the cancer cells, the affected area is illuminated by an intense laser beam at a wavelength strongly absorbed by the dye. This kills the cancer cells that can be reached by the laser light.

Why it should do so is not exactly clear. Dougherty attributes the effect to the molecular fragments produced when the light breaks down the dye. Normal cells absorb less energy because they contain less dye and are able to recover without permanent damage. At this writing, Dougherty's work is in the early stages of experimentation.

HOLOGRAPHIC IMAGING AND ORTHODONTICS

Lasers can be used to record three-dimensional images with a technique called *holography,* which we'll describe in more detail in chapter 12. Holography can be used to record minute changes in three-dimensional objects, such as the effect orthodontic braces have on patients with misaligned teeth.

From an engineering standpoint, the forces acting upon these patients' teeth produce a mechanical deformation within the mouth. The motions produced are often impossible to measure by conventional means. They occur in all three dimensions, with different degrees of motion in each dimension, and vary across the entire surface. Furthermore, the motions are small, sometimes only a few thousandths of a millimeter (less than a thousandth of an inch).

One solution is to record two holograms on a single photographic plate. One hologram shows the mouth before a corrective force is applied, the other afterward, for example, after a wire is tightened on orthodontic braces. The combination of the two holograms shows the deformation across the entire surface and measures it in units of the wavelength of the laser light used to record the hologram. The mechanism by which the measurement is made is explained in more detail in chapter 12. The same principles used in orthodontics can be applied to the treatment of broken bones or bone disease.

A few years ago, the scientific journal *Applied Optics* published a paper in which a group of researchers described how they used holography to record the motion of teeth caused by braces. The research was interesting, but the most memorable part was the photographs illustrating it. The look of misery on the patients' faces was not simply due to their being subjected to braces; the use of holography required that the insides of their mouths be painted with a special paint to record the holograms. Readers facing the prospect of orthodontic work may be relieved to learn that the technique remains in the laboratory stage!

THE STATE OF LASER MEDICINE

Many other things are going on in laser medicine, and in this chapter we have only been able to skim the surface. We don't have room to cover all the experiments with lasers that have been performed on animals, or even humans.

The laser has a bright future in medicine, but a limited one. It offers no miraculous cure for any dread disease. Instead, the laser is a very useful tool that can aid in the treatment and diagnosis of a number of serious diseases. Barring dramatic breakthroughs, which don't seem to be on the horizon, most readers of this book will never be treated with a laser.

With a few exceptions—notably gynecological surgery and the two eye treatments discussed earlier—laser medicine remains a subject of research, rather than a generally accepted technique, although it should be pointed out that much of that research involves clinical experiments performed on human patients.

Also, some of the results are being called into question by other researchers, as in the case of some types of laser dermatology, such as tattoo removal. These questions, in turn, call for more research. It takes time to evaluate medical techniques carefully and make certain there are no harmful side effects.

Even techniques considered effective don't work all the time. For example, 6.4 percent of the eyes treated for diabetic retinopathy with a laser during a National Eye Institute study became functionally blind. Nonetheless, the study strongly recommended the laser treatment, because 16.3 percent of the eyes left untreated also became functionally blind during the same period.

HOPE AND PROGRESS, HERE AND ABROAD

Two decades after laser medical research began, the field remains an exciting one. It's also matured considerably. The pioneers in the field had almost no data to start with. They didn't know what a laser beam would do to the skin or other tissue and to parts of the body. They tried to find out, sometimes taking risks with their own bodies, in experiments that today may not seem very sophisticated. Today's researchers have a wealth of data, accumulated by these pioneers and those who followed them. They also have more, and better, laser equipment to work with.

But there are problems. The elaborate safety regulations on medical instruments in the United States, coupled with the prevalence of malpractice suits, tend to slow the adoption of new laser techniques in this country. They also cut down the number of clinical experiments, in which new techniques are tried on human subjects. Although these restrictions help protect patients from unexpected side effects, they also limit the spread of effective new treatments. As a result, the United States is falling behind Western Europe in laser medicine, or at least in its application.

China is making impressive strides, despite serious limitations. Chinese researchers, who cannot readily purchase lasers, sometimes have to make their own or spend considerable effort finding other people to make lasers for them. Technical skills are still in short suppy, because the Cultural Revolution closed the nation's universities for about a decade. Even so, visiting American and European physicians have found that they can learn much from the Chinese, who have gone further in testing new methods on human patients than have physicians elsewhere in the world.

6 LASER COMMUNICATION: TOWARD THE FIBERED SOCIETY

Long before people learned how to use electricity, long-distance communication was by light: by signal fires, smoke signals, or semaphores on hilltops. The invention of the telegraph did not kill the idea of optical communication. A century ago, Alexander Graham Bell turned his attention to transmitting voices over beams of light. However, it was an earlier invention of Bell's, the telephone, that has set the tone of communications for the past century.

For the last two centuries, people have been developing ways of sending more and more information faster and faster. The telephone transmits voices, which, from the standpoint of the communications scientist, represent more information per unit time than telegraph signals, which carry, at best, a few letters per second. A television signal, in turn, carries more than a thousand times as much information as a telephone line.

It all adds up to a lot of information being transmitted, and that presents a problem. The amount of information that can be carried by an electromagnetic wave depends on its frequency—the higher the frequency, the more information can be transmitted. But the amount of information to be transmitted has increased faster than the available ways to transmit it, creating a bottleneck.

Communication by light offers a way around this problem. Voice frequencies transmitted by telephone are a mere 1,000 to 4,000 cycles per second, or hertz. Television signals have frequencies of around 50 million hertz. The frequencies of light waves, on the otherhand, are around 800 *trillion* hertz.

The theory behind transmitting information by light was around for years before there was a way to put it into practice. The first requirement was a suitable light source, which had to await the development of the laser. The second requirement was a suitable way to transmit the light, which emerged a decade ago in the form of the *optical fiber*, a hair-thin strand of glass that guides light along its length. The combination of laser, fiber-optic, and computer technology could revolutionize the way in which we communicate with other people in the coming decades, by interconnecting homes, businesses, and government to form a "fibered society."

THE OPTICAL TELEGRAPH AND THE PHOTOPHONE

The science of rapid long-distance communications began nearly two centuries ago with the optical telegraph. This was the brainchild of Claude Chappe, a French engineer who was trying to meet his country's need for fast and reliable communications during the turbulent early 1790s. His system was a simple one: a series of hilltop towers, to which vertical posts carrying moveable wooden beams were attached. The beams served as a semaphore; each letter of the alphabet was represented by a different semaphore position. A man sat in each tower and watched for signals from the adjacent towers, which had to be visible. The system eventually grew to over 500 towers stretching over 3,000 miles.

The optical telegraph was labor-intensive, and even in that era of cheap labor, it was expensive to operate. However, it remained by far the best means of communication available until the electrical telegraph replaced it around 1850.

The idea of communication by light was resurrected a quarter of a century later by Alexander Graham Bell, who turned his attention to finding a way to transmit conversations using light shortly after he developed the telephone. In 1880, he demonstrated what he is said to have considered his most important invention—the photophone.

The photophone transmitted voices—and voices only—by using the sound waves to vibrate a thin, flexible mirror. The vibrations altered the direction in which the mirror reflected sunlight, causing some of the light to miss the remote receiver, thereby modulating the light with the voice signal. The hardest part was converting the modulated light into an electrical current (done by a primitive ancestor of today's photocells), so that the photophone could use the same type of earpiece that Bell had used for the telephone.

Bell and his assistant Sumner Tainter performed their first successful photophone experiments on February 19, 1880, in Washington, D.C. More experiments followed, but although Bell refined the hardware, the photophone wasn't practical. Clouds could block the sun, and even with artificial light sources, rain, snow, or fog could block the beams of light.

Shortly after his first experiments, Bell put his original photophone into storage at the Smithsonian Institution. There it remained for nearly a century, virtually forgotten.

In the intervening years, there were a few attempts to use optical communications, but little came of them until lasers and fiber optics arrived on the scene. These two new technologies revived interest in optical communications, particularly at Bell Labs, which has one of the world's largest research programs in the field. As a result, Bell's original photophone was retrieved from storage, and on February 19, 1980, his original experiments were reenacted at the site of his Washington lab (now a parking lot) by representatives of Bell Labs, the National Geographic Society, and the Smithsonian, along with Forrest M. Mims III, a San Marcos, Texas, inventor and writer, who was a driving force behind the photophone centennial. (Ironically, Mims was also involved at the time in a dispute with Bell Labs over rights to an invention.)

ENTER THE LASER

Scientists were quick to realize the potential of the laser for communications. Even before the first laser was demonstrated, people who understood the concept were suggesting its use in communications. Others followed soon after the laser was actually invented in 1960.

By August 1962 Isaac Asimov was writing about the potential of optical communications in his monthly column in *The Magazine of Fantasy and Science Fiction.* He ran through some calculations that indicated that, in theory, the visible portion of the spectrum could be used to transmit up to 100 million television channels, while the ultraviolet region could be used to transmit six billion additional channels. Then he thought about the low quality of television programming: "Imagine what the keen minds of our entertainment industry could do if they realized that they had a hundred million channels into which they could funnel new and undreamed-of varieties of trash. Maybe we ought to stop right now!"

Scientists rarely heed warnings to stop, and the developers of optical communication systems were no exception. They set up lasers in their laboratories and began rediscovering what Alexander Graham Bell had learned eighty years earlier—that the atmosphere isn't a very good medium for transmitting light.

Researchers at Bell Labs soon turned to an idea that goes back to a patent issued in 1934 to Norman R. French—the light pipe. In its simplest form the light pipe is literally that—a pipe with a reflective inner surface. If the surface reflects a large enough fraction of light, a beam shone down the tube will keep on going, although with small losses. More refined versions were soon proposed, such as pipes containing a vacuum to prevent light absorption by the air, and pipes containing a series of lenses to refocus the beam and make certain it follows the desired path.

Meanwhile other researchers looked upward to space, where there is no air to absorb or scatter light. The National Aeronautics and Space Administration studied the prospects for communication from ground to satellites and between satellites. Military planners were interested in the same ideas, largely because the narrow beam from a laser would provide much more secure transmission than radio waves, which radiate in all directions. The hottest current military project is the development of a system that would use blue-green lasers to relay signals from satellites to submerged submarines.

DAWN OF THE FIBER ERA

We ordinarily think of light as traveling in a straight line. It's possible, however, to make thin fibers of glass that can carry light signals around corners and bends, in a way that is superficially similar to the way copper wires carry electricity. These optical fibers rely on a phenomenon known as total internal reflection (see diagram 7).

This means that when light traveling through a dense material (say, glass) strikes an interface with a less dense material (say, air) at a low, glancing angle, all of the light can be reflected back into the denser

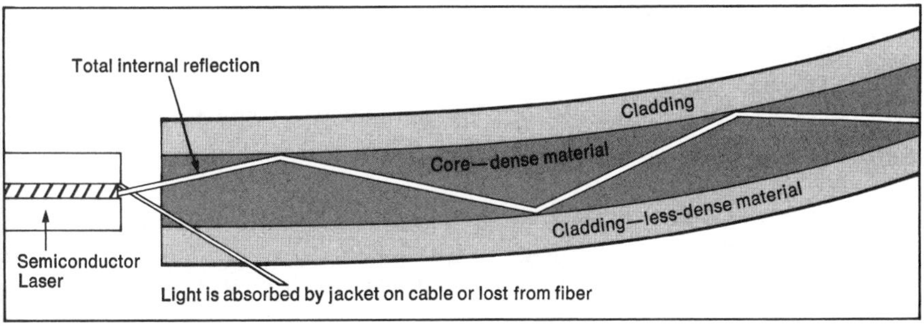

7. Optical fibers rely on a principle called *total internal reflection.* All this means is that when light traveling through a dense material strikes the interface with a less dense material at a small angle, as shown, all of the light can be reflected back into the denser medium. In practice, an optical fiber is made of two types of glass: a dense glass for the core and a less dense glass for the outer cladding. If light in the core hits the interface with the cladding at a small enough angle, it will bounce back into the denser core and keep on going through the fiber, even around bends. If the angle is too large, however, the light passes through the cladding and is absorbed by the jacket on the cable or is simply lost from the fiber.

medium. What this means in practice is that you can make an optical fiber out of two types of glass: a dense glass for the core, and a less dense glass for the outer cladding. The light will travel in straight lines through the core until it hits the interface with the cladding. If the light hits at a gentle enough angle, it will bounce back into the denser core and keep on going through the fiber, even around bends.

A century ago, British physicist John Tyndall demonstrated the concept with water flowing out of a tank. In the 1950s, the American Optical Corporation developed light-transmitting fibers made of two types of glass. These early fibers could carry light around corners, but not very far—no more than tens of meters (or yards).

It wasn't until 1970 that the first long-distance fiber was made, by Robert Maurer at the Corning Glass Works. Maurer was inspired by research into the absorption of light by ultrapure glass by Charles Kao and George Hockham at Standard Telecommunication Laboratories, a British subsidiary of the International Telephone and Telegraph Corporation. In 1966 Kao and Hockham predicted that glass fibers should be able to transmit light well over a kilometer (or even a mile). As it turned out, their optimism was well justified.

Some numbers will give you an idea of how far the technology has come. In the mid-1960s, the best commercial fibers absorbed about 90 percent of the light that entered them within only 10 m (about 11 yd). This distance was increased to 500 m (about 550 yd) with Maurer's first low-loss fibers. Today, the best fiber fabricated in a laboratory can transmit light for well over 30 km (about 20 miles) before 90 percent of the input light is absorbed. All these distances are for the wavelengths where absorption is the lowest—in current fibers, about 1.3 to 1.5 micrometers, in the infrared region, about twice the wavelength of visible light.

PRINCIPLES OF FIBER OPTICS

Fiber-optic communication is somewhat analogous to electronic communication, except that the signal takes the form of light. A light source, or transmitter, generates light that has had a signal superimposed on it through a process called *modulation.* The light signal is transmitted through an optical fiber, which, in practice, is encased in a cable that from the outside looks like an ordinary (though small in diameter) electrical cable. At the end of the fiber is a receiver that "catches," or detects, the optical signal, decodes the information (*demodulates* it from the light), and amplifies the resulting electrical signal for the next stage in the communications network.

Two types of light sources are used. One is the semiconductor laser: a crystal no bigger than a grain of salt that contains a tiny laser resonator and emits a beam that is recognizably a laser beam, although it spreads out more rapidly than the beam from other kinds of lasers. The other is the *light-emitting diode,* or *LED,* a semiconductor crystal that is similar to a semiconductor laser except that it doesn't contain a laser resonator and emits a broader beam. (Most fiber-optic light sources actually emit infrared, and strictly speaking are *IREDs—infrared-emitting diodes*—but we'll go along with the vast majority of engineers who just call them LEDs anyway.)

When an electrical current that exceeds a certain *threshold* value is passed through a laser or an LED, the semiconductor chip emits light. Once you exceed the threshold current, the amount of light is proportional to the current passing through the chip: increase the current and you increase the light. Thus, variations in the current show up as

The small rectangle with wire attached on top of the polished block is a diode laser. This is the type of laser that provides light for fiber-optic communications. The white block to the right is a grain of salt, which gives you an idea of how small diode lasers are. Courtesy Bell Labs

variations in the light—or modulation of the beam. So the light beam can carry the same information as the electrical signal driving the laser or the LED. The modulated beam travels through the fiber and is caught by a receiver at the other end. In the receiver, a semiconductor device called a *detector* produces an electrical signal proportional to the amount of light that strikes it.

From the standpoint of the receiver, it doesn't matter if the light source is a laser or an LED. In fact, that distinction is important only to the engineer designing a fiber-optic system. For that reason, we're going to change our focus for the rest of this chapter and talk about fiber optics rather than about lasers *per se.* Although lasers are not used in every single fiber-optic communication system, fiber optics is an exciting technology whose growth has been stimulated by the laser.

Lasers are likely to always play a role in some fiber-optic communications, but how large that role will be is unclear. LEDs are giving lasers a run for the money, because LEDs are less expensive and last longer. The main advantages of lasers are that they emit more light and can be modulated faster. At the moment, a rough rule of thumb is that lasers are used to transmit information rapidly over distances of 1 km (0.6 mile) or more, whereas LEDs are used for transmission at lower speeds and over shorter distances. Advances in LED and laser technology may change that, but you'd never notice the difference.

SQUEEZING IN INFORMATION

If fiber optics were only able to do as much as electronic communications, there wouldn't be much use for fibers. But they can do much more. Fibers can break the information bottleneck, because they can transmit much more information in a given amount of time than wires. A single fiber, for example, can carry the equivalent of over a thousand telephone conversations.

This greater capacity is critical. Everything that is communicated can be looked upon as information, whether it's music, a television program, a series of telegraphic dots and dashes, or digital data for a computer. The rate at which information can be transmitted has traditionally been a major limitation on communications.

Fiber optics may be the answer. The difference between wires and fiber optics is dramatic. Ordinary telephone conversations in *analog format* (see box) have a *bandwidth* of 3,000 cycles per second, or hertz. (Bandwidth, measured in hertz, is one indication of how much information can be transmitted). High-fidelity music has a bandwidth of about 20,000 hertz. Television signals have a bandwidth of about 6 million hertz (6 megahertz). The transmission capacity of optical fibers, on the other hand, has now surpassed 1 *billion* hertz in laboratory demonstrations.

These values are all for the analog format, which is used in most existing communications equipment. Optical fibers themselves operate quite happily in analog format. However, there is a problem with lasers—they distort analog signals. And LEDs have a different limitation—they can't operate as rapidly as lasers and hence can't transmit information as fast.

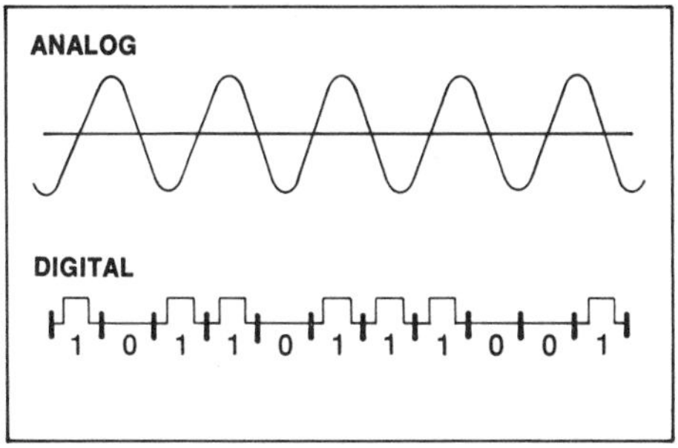

ANALOG AND DIGITAL COMMUNICATIONS

Information can be transmitted in either of two formats: *analog* or *digital.* An analog signal is one that varies in amplitude continuously, such as sound. Almost all home electronics that reproduce sound or television signals use analog format.

In digital transmission, information is encoded as a series of zeroes and ones (see diagram). Digital format was originally developed for numerical information, and pocket calculators and computers use it.

Any information in one format can be translated into the other and back again. This can be done repeatedly, and it sometimes happens in the telephone network. The need for conversion arises because rapid advances in digital electronics have pushed it far beyond older analog electronics; thus, the new digital equipment that telephone companies are installing must be compatible with the billions of dollars worth of existing analog hardware. Besides offering higher-fidelity reproduction (as in many modern sound studios), the digital format permits computer processing and switching of information—a feature particularly important in telecommunications.

Fortunately, there's a way around this problem—shifting to *digital* transmission, in which information is encoded as a series of zeroes and ones (see box). These *bits* of information can be electronically decoded to generate a signal in analog format, such as high-fidelity music. Lasers, LEDs, fibers, and detectors all handle digital signals without difficulty.

LONG-DISTANCE TRANSMISSION

There's a second major advantage to fiber optics. Not only do fibers transmit information faster than wires, they can also transmit it farther. The key breakthrough that made fiber-optic communications possible was a dramatic reduction in the amount of signal lost in the fiber—corresponding to a dramatic increase in the transparency of the glass used. Researchers have kept on going. Now fibers that transmit signals much farther than ordinary wires are mass-produced.

To understand the importance of this breakthrough, let's look briefly at how a long-distance communication system works. A signal is transmitted as far as it will go along a wire, optical fiber, or some other transmission medium. The low-level signal at the end of the line is then amplified to produce a higher-level signal that can be fed into another length of transmission line. The thing that does the amplifying is called a *repeater.*

A repeater may sound simple, but in practice it can be a lot of trouble. As well as amplifying the low-level signal, a repeater has to make sure that the output signal looks like the signal fed into the original transmission line. Thus, repeaters are expensive to build, and they also have to be installed, often outdoors and sometimes in hard-to-reach locations. What's more, things can go wrong with them, requiring repairs that can be very difficult and costly if the repeaters are hard to reach.

Because of these problems, operators of communication systems would like to reduce the number of repeaters they use. Fiber optics can be a big help, particularly for the operators of the most pervasive communication system, the telephone network.

Telephone companies are most interested in fiber optics for the "trunk" lines that carry many telephone conversations simultaneously between central offices. Metal-cable trunk lines connecting central offices require a repeater about every 1.6 km (1 mile) or so. Fiber-optic systems can easily operate with repeater spacings of 6 km (4 miles) or more—equivalent to the average spacing between central offices in urban and suburban areas. That's important, because a telephone office is a far more benign environment for a repeater than the bottom of a manhole or many of the other places where repeaters are housed. What's more, new developments in fiber optics are stretching repeater spacing up to as much as 20 km (12.5 miles).

FIBERS UNDER THE WATER

A more dramatic—but further-off—use of fibers is in undersea cables. Although satellite circuits are convenient, it costs about three times as much to make a telephone call using a satellite as it does over trans-Atlantic cable, according to Peter Runge, a Bell Labs scientist. Besides, the time required for a signal to go from the earth to a satellite and back is long enough to be perceived, and this delay is annoying during a telephone conversation.

The Bell System is planning a new trans-Atlantic cable in the late 1980s, and current plans call for the cable to contain optical fibers instead of wires. More information—the equivalent of 4,000 telephone calls—can be transmitted over each fiber than over a single circuit in a conventional metal cable, and the spacing between repeaters can be increased. This is a great advantage, since a malfunctioning repeater at the bottom of the ocean is no minor problem. Fixing it could require hauling up hundreds of miles of cable—a process that in itself could mechanically damage the cable and thus create further problems. Because recovery of cable is expensive and time-consuming, engineers go to great lengths to design long-lived repeaters.

Metal undersea cables require one repeater every 9 km (about 6 miles). Bell Labs' goal for its fiber-optic cable is a repeater spacing of 35 km (about 22 miles), meaning that only about a quarter as many repeaters would be needed. Repeater spacings as long as 100 km (about 60 miles) might be possible with the best fibers made in the laboratory, but Bell System engineers don't want to cut their margins too thin.

There's another advantage to fiber-optic cables: their small size would make them easier to install than conventional, metal undersea

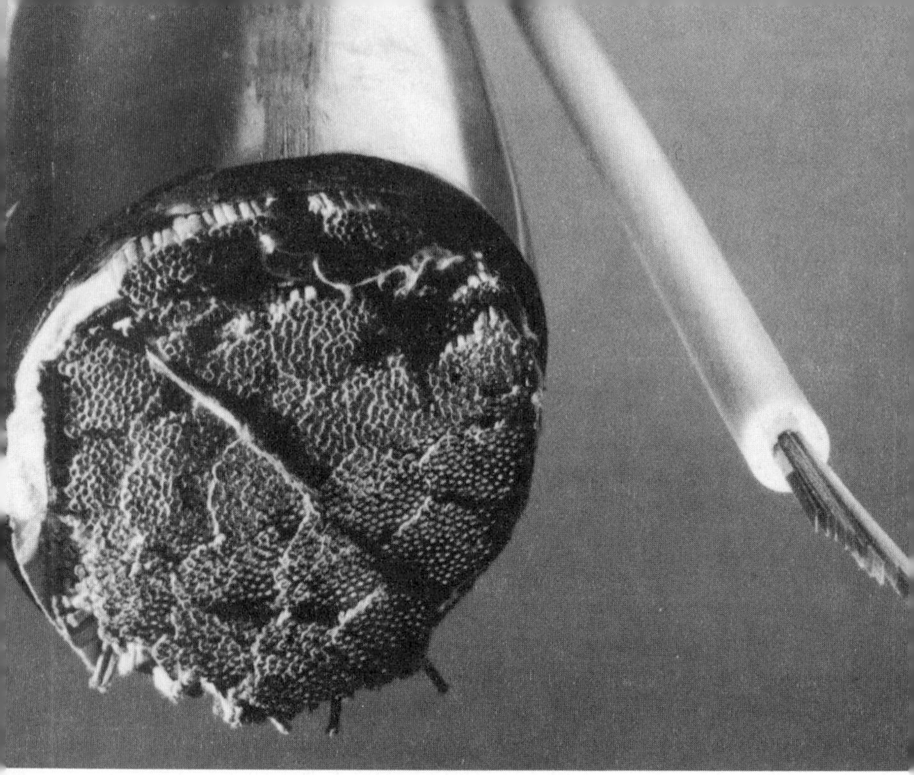

To the left is a conventional wire telephone cable. To the right is a fiber-optic cable, with individual glass fibers bonded together in ribbons and then encased in a protective sheath. Despite being much smaller, the glass cable carries nearly 50,000 simultaneous conversations—more than three times as many as the larger wire cable. Courtesy Bell Labs

cables. It's no wonder that undersea fiber-optic cables are being developed around the world—in Japan, Britain, and France, as well as the United States.

THE FIBERED AUTOMOBILE

There's yet another advantage to fiber-optic communications. Glass fibers carry signals in the form of light rather than electricity, so there are no sparks that could cause explosions in chemical plants, oil refineries, or other sorts of volatile environments. This also means that fiber optics won't pick up signal-obscuring "noise" caused by nearby electrical currents—the equivalent of the static you pick up on your AM radio when you plug in an electric shaver or vacuum cleaner nearby.

Avoiding harmful effects of electrical noise is a major concern of today's automotive engineers. Microcomputers and other electronic components are being used more and more in car controls, and the very small signals required by modern semiconductors can easily be obscured by the electrical noise generated by others parts, notably spark plugs. General Motors is considering fiber optics as a way around this problem.

Other considerations favoring fiber optics are also stressed by Wesley A. Rogers, former manager of advanced electro-optics at the General Motors Technical Center in Warren, Michigan: the network of wires in today's cars is both heavy and difficult to service, and, since virtually all of the controls are located on or near the dashboard, all the wires must pass through that region. This jumble of wires next to the controls is a pain in the neck for auto mechanics.

Rogers's alternative is to carry the control signals over optical fibers from the dashboard to other parts of the car. Power wires would go directly from the battery to fiber-operated controls, which would do such things as tell a power-window motor to roll the window down. The result would be a simpler—and much lighter—wiring harness, which would help automakers meet fuel-economy goals. GM also expects the fiber-optic system to be simpler to service than conventional wiring networks, which the authors can testify are difficult at best.

The fiber-optics technology that GM is seeking is poles apart from what Bell Labs has developed for telecommunications. Where Bell Labs speaks of alignments accurate to thousandths of a millimeter, GM thinks of idiot-proofing assembly and service and alignment requirements that are much less stringent. Where Bell is developing sophisticated technology, GM is working on cheap but durable components, such as LED light sources instead of lasers. And where Bell wants to transmit tens of millions of bits of information per second, GM plans to transmit only 1,000 control signals each second.

Fibers are already being used to transmit light from bulbs in some turn signals and other automobile indicators. But don't expect to see fiber-optic communications in cars soon. Nor will fibers be used initially in such vital systems as windshield wipers. Instead, they'll first appear in controls for such accessories as power windows, whose failure doesn't make a car unsafe. Because automakers introduce innovations from the top of the line down, you should expect to see fiber-optic communications appear first in Cadillacs, probably around the 1985 model.

FIBERS AND YOU

Uses of fiber optics may still seem remote to you. Actually, communications fibers are as close to you as your television or telephone. If you watched the 1980 Winter Olympics or a football game from Tampa Bay Stadium, you've already seen laser–fiber optic communications in action—transmitting television signals.

At Lake Placid, New York Telephone installed a temporary fiber-optic link between the arena where the Winter Olympics were held and the television control center being used by broadcasters. The signals were transmitted optically only as far as the television facility; from there they were transmitted around the nation by microwaves and other more traditional means. The results were excellent, and Pacific Telephone plans to use fiber optics in televising the 1984 Summer Olympics in Los Angeles.

In Tampa, General Telephone installed a fiber-optic cable to transmit television signals from the football stadium to another point in Tampa where they could enter the national television system. The Lake Placid system was installed above ground, where it had to face the harsh Upstate New York winter, but the Tampa link was installed in permanent underground ducts. Although the environment in Florida may sound more benign, manholes there are continually damp, which can degrade communications equipment, and are sometimes inhabited by wildlife—including an occasional alligator.

Fiber optics are also being put to use elsewhere in the television industry. Fibers are used in studios because of the high quality of transmission. Fibers are used to carry signals from some remote television cameras, particularly for news and sporting events, because they are light and compact. Fibers can also transmit television signals over very long distances.

The world's longest fiber-optic system under construction is 3,200 km (2,000 miles) long and will bring cable-television services to the fifty largest towns in the Canadian province of Saskatchewan. (Fibers turn out not to be economical for connecting individual homes to cable-television networks at the moment. But that will probably change when new communication services are offered.)

Your telephone calls, too, may pass through fiber-optic cable without your being aware of it. The Bell System and other telephone companies are using fibers to provide the high transmission capacities required between busy central offices. The biggest telephone system in the works will run from Cambridge, Massachusetts, to Washington, D.C., along the busy Northeast Corridor, which carries the nation's heaviest concentration of telephone traffic. That system will be able to carry tens of thousands of telephone conversations at once at any single point. A similar system connecting Sacramento and Los Angeles is in the planning stages.

Fiber-optic connections are quietly being made in dozens of places in the country's telephone network: in rural north-central Pennsylvania and in Sugar Land, Texas; in Chicago and Las Vegas; in downtown Manhattan and across the Golden Gate Bridge. The same thing is going on in Argentina, Canada, Poland, Italy, Japan, England, France, and elsewhere.

The impact of all these fiber-optic installations on you will be indirect, primarily in the form of lower costs and better service. They're also a quiet harbinger of things to come.

TOWARD THE FIBERED SOCIETY

In ten or twenty years it should be possible to bring the broadband capabilities of fiber optics all the way to your home, as fiber-optic technology develops and becomes less expensive. The fibered society will make possible all the communications services envisioned a decade or more ago for "wired cities"—as well as services not even dreamed of then.

Some writers glibly toss the term "wired society" around as if it were something in the future. The fact is that we already *are* a wired society, wired with a nationwide telephone network, which is interconnected with telephone networks in other countries, so that you can talk with people halfway around the world. That same network can carry digital data to and from computers or transmit coded signals that can be used to generate a facsimile of an original document in a few minutes.

Parts of this country are also wired with cable-television networks.

There's an important difference between cable-television and telephone systems. The telephone network allows any two individuals to talk to each other, and to each other only (if the hardware is working properly). In a cable-television network, *all* channels are transmitted from a central site to each subscriber, who then selects from among them.

The key concept of the fibered society is creation of a video network like our present telephone system, which could make connections for transmission of televisionlike signals between any two subscribers or, more important, between a subscriber and a provider of specialized information. This means that an individual subscriber could order video German lessons, stock market reports, movies, or almost anything that can be put on a video screen, so long as the provider has it in his library or computer memory.

The Department of Housing and Urban Development (HUD) tried to test the wired-city concept around 1970, using coaxial cable. The hardware was relatively primitive (there wasn't much ability to choose among information providers), but even so, it was expensive. The technological limitations were compounded by political problems—the demonstrations were part of the "new towns" program, which fell on political hard times during the Nixon Administration.

The first person to suggest using optical fibers as the basis for a wired city was John Fulenwider, then at GTE Laboratories and now at Arthur D. Little, Inc. With the HUD program winding down, his 1972 suggestion fell on deaf ears in the United States. But it was not ignored elsewhere.

JAPAN'S FIBERED CITY

Over the past decade, Japan has made tremendous strides in commercial electronics, particularly in consumer products. The progress has been so rapid that many U.S. electronics companies are running scared. Since about 1975, Japan has been making a similar push in fiber optics and, as a result, has become the world leader in many areas of optical-fiber and semiconductor-laser technology. Japan was also the first to demonstrate broadband fiber-optic communications to homes, on July 18, 1978, in Higashi Ikoma. At this writing, that experiment is continuing.

It's called HI-OVIS, for Highly Interactive Optical Visual Information System, and serves 158 homes. Its exact cost hasn't been disclosed, but around 1980 knowledgeable estimates ran in the $25- to $40-million range. Most of the money has come from Japan's powerful Ministry for International Trade and Industry.

The goals for HI-OVIS include evaluating fiber-optics technology and helping Japan develop a fiber-optics industry. But the most interesting part of the experiment involves the four major *social* goals of the program:

- Establishing a new community of people linked together by HI-OVIS—a sort of video neighborhood.
- Making educational courses available in the home.
- Developing a "safe local-welfare society" by encouraging people to help each other and by helping deliver medical, police, and fire-protection services over the communications network.
- Making subscribers actively select a source of information rather than passively watch broadcast television.

The central element of this social experiment is an interactive television channel. Technically, the transmission originates at a studio in HI-OVIS headquarters. However, the subscribers have television cameras and microphones in their homes, which they can use to send homemade programs to the studio—much in the way people use telephones to participate in radio talk shows. Engineers at the studio can select a signal from the studio or one from a subscriber's home to send to all homes watching the program.

Programs on the interactive channel are designed to encourage participation. Some programs feature discussions of political issues. In others, subscribers share experiences. One example is a cooking program that encourages viewers to show others how they make their favorite recipes in their own kitchens.

HI-OVIS also functions as a cable-television system, retransmitting television signals from remote stations to subscribers' homes. Subscribers can use their home control consoles to request that one of a handful of videotapes in a small library be played especially for them. That library once included children's programs, but the children became such enthusiastic users of the service that no one else could use it—and in the process they wore out some of the videotapes by playing them more than 500 times. Viewers can also request displays of still pictures, including frequently updated information, such as train and plane schedules.

The Japanese seem to be satisfied with the initial results. "HI-OVIS has become a part of the lives of [subscribers]," a preliminary report concludes. The next step under consideration is expansion to create HI-OVIS II, which would serve a few thousand homes and probably cost over $100 million. If it gets the go-ahead HI-OVIS II could be in operation by 1985.

FIBERED FARMS—THE CANADIAN VERSION

A somewhat less ambitious—and much less expensive—test of fiber-optic service to homes was to begin in rural Manitoba in September 1981. This system is to serve about 150 homes: 50 each in the villages of Elie and St. Eustache and 50 farms between the two. It's intended as a step toward Canada's goal of providing rural communities with telecommunication services equal to, or better than, those enjoyed by urban residents, according to Brian B. McCallum, who helped organize the test. Canada's interest in improving the quality of rural life reflects the importance of agriculture to the country's economy.

The cost is about $6.3 million, only about one-fifth that of HI-OVIS, but the services are similar except for the absence of an interactive channel. The Manitoba subscribers can pick two channels at a time out of nine available; each HI-OVIS subscriber can receive only one channel at a time (meaning that all television sets in the house must be tuned in to that one channel). Canadian subscribers can choose from among seven stereophonic music signals, a service not available from HI-OVIS. The optical fibers also provide telephone service in the Canadian system (but not in the Japanese system); many of the Manitoba subscribers are getting single-party telephone service for the first time in their lives. In fact, some of them only recently had the number of parties sharing their telephone lines reduced to four. The Manitoba subscribers also have a channel for communicating with a computer, but they cannot request specific video programs from a stored library.

FRANCE'S VISION OF THE FUTURE

At this writing, the latest and most ambitious vision of the fibered future to be translated into hardware is in France. The French Directorate of Telecommunications, which operates the country's telecommunications network, is spending about $100 million to install a fiber-optic system connecting some 4,500 homes in the resort city of Biarritz. Current plans call for service to the first 1,500 homes to begin in early 1983. The system is part of an effort to push technological development in France.

Biarritz is the first project to include videophone service. The idea is similar to the Bell System's Picturephone, which never really got off the ground in the United States. However, the French system may have a better chance, because its technology is more advanced. It uses the full resolution of a European television screen, which has about 20 percent more lines than a U.S. screen, in contrast to the much lower resolution used in Picturephone. Because the videophone service is not expected to be much more expensive than regular French telephone service, planners of the Biarritz system hope to avoid another deterrent to Picturephone use—its high cost.

Each Biarritz subscriber will receive a color television set and a black-and-white television camera. Although the videophone service can operate with color-television cameras, there aren't any French-made color-television cameras that can compensate for home-to-home variations in lighting. Rather than importing cameras, the French government decided to supply French-made black-and-white cameras only. Users can add their own color-television cameras if they wish.

In addition to videophones, the system will provide a choice among five color-television channels, three in French and two in Spanish (because Biarritz is near the Spanish border). Videotapes and local broadcasts will be added to make at least fifteen television channels available. One possibility might be continuous monitoring of the surf on the resort city's beaches with a remote television camera, according to Alain Bernard, Directorate engineer responsible for the project. Subscribers will also have their choice of five channels of stereophonic music.

FIBERS AND THE FUTURE

HI-OVIS, Elie, and Biarritz barely scratch the surface of the possibilities offered by broadband fiber-optic networks. Development of fiber optics and computer technology will make it possible to tie the world together with a single integrated network for communications and information retrieval by the turn of the century. The first parts of that system could be put in place by forward-looking countries such as France, Japan, or Canada before 1990. (Although Bell and other telephone companies are effectively putting fiber optics to work in the telephone network of the U.S., present government policies make it impossible for any single organization to push integrated broadband communications in most of the country. As a result, the U.S. is far behind.)

There are a few technological barriers remaining, primarily the lack of practical methods for switching optical signals. But given the amazing pace of fiber-optics research, those barriers are unlikely to stand for long. The first phase of HI-OVIS was obsolete almost as soon as it went into service. In the first part of this decade, systems are likely to become obsolete in the time required to translate concepts into operating networks, muses Brian B. McCallum, who helped plan the Elie system.

The future of telecommunications is beyond the scope of this book. However, we'll close this chapter by giving you a taste of some of the wonders to come from the integration of fiber-optics and computer technology:

• *Home work stations* with video terminals and keyboards. You would be able to work at home, communicating by videophone with fellow employees at their home work stations and retrieving information from computerized data banks maintained by your employer and other organizations. The work station might include a television camera and a facsimile transmitter-receiver capable of reproducing copies of paper documents through the communication network. It would almost certainly be able to talk to you; it might even be able to understand your spoken words.

• *Information retrieval* from libraries all over the world. These libraries would include not just printed texts but also audio and video programs. If, for example, you wanted to watch "Gone With the Wind," you could have a video library transmit the movie to your home (a service for which you'd pay a fee). Books won't vanish, however—at least not until someone comes up with a display device that's more portable and easier to read than a cathode-ray (television) tube.

• *Home monitoring services,* such as fire and burglar alarms, automatic thermostat control, meter reading, and the like.

• *Banking and shopping* by telecommunications from your home.

• *Immediate translation* of an ongoing conversation. You could speak in English to your home work station, which would recognize your words, then transmit them to a translation computer. That computer could translate them into another language—say, Japanese—and transmit the translation to another work station halfway around the world. That work station could synthesize a voice to speak your words—in Japanese. The system would be nowhere near as compact as the hand-held automatic translators that science-fiction writers have used for years to avoid language problems in their stories. And the translation wouldn't be elegant. But the system should work if some sticky problems in computer recognition of speech can be worked out—most likely by programming the computer to recognize only one person's speech.

7 DEATH RAYS . . . AND OTHER WEAPONS

You've probably heard the stories:

• That the Russians, with a massive laser situated on a mountaintop, have knocked out one of our spy satellites.

• That the U.S. Army, in tests as early as 1976, shot down remote-controlled drone missiles, and that our Air Force has a giant laser that occupies an entire plane, which it intends to use in conventional air attacks.

• That both the Soviets and the American military are planning laser-armed space stations, and that World War III will be fought in orbit, á la *Star Wars.*

• That Russia is far ahead of the U.S. in developing killer lasers, and that we have to close this "laser gap."

There is *some* truth to all of these stories. Perhaps, though, you've also heard the following tales:

• That two respected scientists, one from the Massachusetts Institute of Technology, the other from Carnegie-Mellon University, have pooh-poohed the whole idea of space-borne laser weapons, saying they'll never deliver enough energy to their targets to destroy them.

• That some observers of the military/Washington scene feel that reports of Soviet laser prowess are simply false "leaks" to the press—intended to convince the public and Congress of the need to spend billions of dollars to escalate our own laser weapon programs.

• That one of the inventors of the laser feels that military lasers are getting so big that there's only one way to use them effectively: drop them on the enemy!

And there's *some* truth to these statements also. The subject of military lasers is a complicated one. Are lasers capable of shooting missiles out of the sky or waging war in space? We don't know yet. The people who have come down strongly on one side or the other are generally giving the answers they want to hear. But one thing is certain: military lasers are already a reality, and the U.S. and the Soviet Union are trying like hell to build bigger and better ones. This is the story of their attempts and of some of the obstacles standing in the way.

WEAPON OF THE FUTURE

The military is excited about lasers, and with good reason. Already, low-energy lasers have been used to improve the accuracy with which weapons can be delivered to their targets. In fact, these "aiming lasers" have been "the *single most successful* [Department of Defense] investment in the last decade." That's what George Gamota, assistant for research to the deputy under secretary of defense for research and engineering, told the Senate Subcommittee on Science, Technology and Space at the end of 1979.

For the 1980s, the military is focusing on high-energy lasers that wouldn't guide weapons to their targets; they would *be* the weapons. If high-energy laser weapons can be developed—and there's no guarantee that they can—they could dramatically change the rules of defense strategy. Laser weapons might provide a truly effective defense against attack by nuclear-armed intercontinental ballistic missiles. Although that could prove to be good news in the long run, some observers are afraid that in the short run, it could unsettle the uneasy status quo based on "mutually assured destruction (MAD)"—the fact that with present technology, the United States and the Soviet Union would be sure to destroy each other in the event of an all-out nuclear war. In other words, a laser defense system might make nuclear war "thinkable."

The U.S. Department of Defense (DOD) has already spent over $1.5 billion to develop high-energy lasers, and there's more coming. The U.S. *has* shot down simple missiles and drones with high-energy lasers. A much smaller, but parallel, program is looking at the possibility of developing weapons that would disable their targets with beams of charged particles—electrons and/or *ions* (atoms from which one or more electrons have been removed). Together the two efforts are

The Airborne Laser Laboratory, seen here in flight, is a laser with wings. A high-power, carbon dioxide gasdynamic laser and associated equipment occupy most of this converted KC-135 aircraft (the military version of a Boeing 707). Courtesy Department of the Air Force

lumped together as "directed-energy" weapons. We'll just look at the laser side.

The Soviet Union has its own program, but it's not clear how far they've gotten. They're generally credited with having illuminated an American spy satellite over Russia with a laser, *but* the rumors that the satellite was disabled are apparently false; it may have been temporarily blinded, but there was no permanent damage. There are also reports that a Russian laser was used in the war between China and Vietnam to blind Chinese soldiers, which, by the way, would be the only practical way for a laser to be used directly against human beings. The Soviet program appears to be larger than the American, and unlike the U.S., Russia appears to be trying to develop specific laser weapons systems, a move the U.S. thinks is premature. So far the American military is only doing research to see if laser weapons are practical, rather than committing itself to building finished weapons.

In any case, lasers aren't likely to appear on the battlefield immediately. Demonstrating a "kill" in a test is far different from doing so in the heat of battle, as many engineers and soldiers have learned the hard way with other new weapons. "I don't know of anyone who thinks we could build a prototype demonstration [laser weapon] unit," Colonel Frederick S. Holmes, Jr., military assistant to the Pentagon's director of directed-energy technology, said in late 1980. "We haven't been able to demonstrate the range, reliability or maintainability required for military use."

WHAT A LASER WEAPON WOULD DO

Although laser weapons may become a reality within a decade or two, our visions of them are still clouded by the writings of generations of science-fiction writers. In many ways, one of the earliest visions—the Martian heat ray in H. G. Wells's *The War of the Worlds*—is, as we said in the opening sentences of this book, one of the most chillingly accurate. Wells wrote of an "invisible, inevitable sword of heat," which destroyed its targets by heating them until they caught fire. His vision was of an intense beam of infrared light, the invisible form of electromagnetic radiation given off by hot objects. The high-energy lasers furthest along in military development all emit infrared beams, and one of the ways they can destroy targets is by heating them enough to damage critical components (which, as we'll explain below, doesn't necessarily mean burning them up).

The word *destroy* might mislead you to think that the target is vaporized or blown up, but that's not essential. A spy satellite, for example, is useless without its electronic "eyes"—its cameras, or imaging equipment. If those eyes can be blinded—burned out by a laser beam too intense for them to handle—the satellite is functionally dead, even if it remains in orbit. As we have already mentioned, blinding is also the only likely rational use of high-energy lasers against soldiers. Just as in spy satellites, the parts of the human body most vulnerable to lasers are the eyes. Lasers used for this purpose are sometimes given the repulsive name of "eye-poppers."

Such applications may sound dramatic, but they aren't very demanding of high-energy lasers. Eyes, human or electronic, are made to be sensitive to light. Their job is to detect faint signals, and they can be overloaded easily. A laser beam intense enough to damage a human eye might not even be felt on the skin; a beam strong enough to burn out an electronic eye might not make a mark on the metal around it. Remember, you can partially blind yourself by staring at the sun.

It takes much more laser power to damage metal and other harder parts of a target. You have to consider more than just power, though. The way laser energy is transferred, or *coupled,* from the beam to the target is also important.

The first part of the process is simple heating. Any target will absorb some of the laser energy and reflect the rest. The energy that is absorbed goes into heating the target, and as the target's temperature rises, it generally absorbs even more of the incident laser energy. If you zap the target hard enough and long enough, the surface will melt, then evaporate, producing a cloud of ionized gas, called a plasma. Under the proper conditions, the laser beam continues to transfer energy to the target through the plasma and will gradually burn a hole through the target. "Gradually" is a relative term; the whole process takes a matter of seconds if the laser power is high enough.

Simple heating, however, isn't the most efficient way to use a laser. Using a series of high-power pulses, instead of a continuous beam, the laser can cause mechanical, as well as thermal (heat), damage. Each pulse evaporates a little material from the surface. This evaporation generates a shock wave, which weakens the material essentially by pounding on it. A series of closely spaced pulses is particularly effective, because it heats a material while pounding it, and heating makes most metals less resistant to impact.

This combination of thermal and mechanical damage is important, because many military targets are made of aluminum sheets. Aluminum is a hard material to attack because it is so shiny. Even untreated aluminum reflects over 90 percent of the incident light from a carbon dioxide laser, and highy polished aluminum reflects even more. If you try to bore through a thin sheet of aluminum with a laser delivering a power of 100,000 to 1 million watts per square centimeter (about 15,000 to 150,000 watts per square inch), it would take some 30 pulses to melt the sheet. If you tried to break the same sheet by purely mechanical stress (by letting the aluminum cool between laser pulses), you'd need 10 pulses. But applying thermomechanical stress, a combination of the two, it would require only 4 or 5 pulses.

The goal is to damage a critical portion of the target. Burning through a fuel tank might cause the target to burn or explode, or at the very least, to lose its power. Once the skin of a missile or airplane has been penetrated, it may be possible to damage critical internal components, such as electronic circuits, which could be disabled by temperatures that would not harm aluminum sheets. A nuclear warhead could be disabled by destroying some of its critical components. In each case, the target would be "killed" by making it unable to perform its mission, a task that would not necessarily require blowing it up. Indeed, this raises one of the subtle problems of using laser weapons. How do you verify that the target has been killed without blowing it up?

In theory, you could vaporize an entire target with a laser. But why bother? Burning a hole requires only that the entire energy of the beam be focused on one spot. Vaporizing a whole target would take thousands of times more energy, and it's going to be a long time before *that* much laser power is available.

Naturally, the enemy can defend its planes and missiles against lasers. This can be done, of course, with mirrors. Though there's no such thing as a perfectly reflective surface, it is possible to cover targets with materials that will reflect large fractions of incident laser light. Another approach is to minimize the amount of beam energy absorbed by controlling the manner in which material evaporates from the target. Military researchers are working on a variety of such countermeasures, as well as counter-countermeasures to overcome enemy countermeasures. And so forth and so on.

RAY GUNS AND OTHER MYTHS

In the years since H. G. Wells wrote *The War of the Worlds,* generations of science-fiction writers have come up with many visions of much less plausible weapons. Some favorites include the death ray, which kills on contact with any part of the body; the ray gun, which can be adjusted to inflict anything from temporary paralysis to instant vaporization; the disintegration ray, which makes an entire object vanish or turn into a pile of dust by touching any part of it; and the paralyzer ray, a humane science-fiction weapon that simply stuns its victims. These weapons are almost invariably compact enough for the hero to hold easily in one hand. Such weapons have absolutely nothing to do with real-world lasers, which can do none of the above to anything much larger than a gnat.

An X-ray laser might sound like a potential death ray, but there are massive practical problems. It takes a *lot* of energy to make an X-ray laser work; the only known X-ray laser was energized by a nuclear explosion! That's not the type of thing you'd want in your hand. Besides, X rays don't kill people very fast, and on the battlefield the military want quick kills. So much for the death ray.

What then of hand-held "laser guns," the recycled ray guns that science-fiction writers invented as soon as they first heard of the laser in the 1960s? Lots of problems here, too. One problem is that lasers aren't very efficient, and more than half of the energy used to excite the laser medium would end up heating the gun instead. Besides, you just can't get enough energy out of a handgun-sized laser to do as much damage to a person as a bullet can do. Much of the damage inflicted by a bullet is a function of its *momentum* rather than the amount of energy it carries. The momentum is what lets a projectile ram through tissue and destroy it. Photons don't carry enough momentum to do this; a laser beam would have to burn straight through to a critical organ.

We did some back-of-the-envelope calculations to see how much damage a hand-held laser could do. One of the most powerful lasers in the world is the Antares fusion laser at Los Alamos National Laboratory. It produces about 50 joules of energy per liter (about a quart) of gas. Assuming a laser gun could hold a liter of gas, produce 50 joules in a pulse, and focus the beam onto a one-square-centimeter (roughly 0.15 sq in) area of your enemy's bare skin, the best you could do is burn him. Your laser gun probably wouldn't be able to burn through his clothing, although it might get hot enough to burn *your* hand. It certainly wouldn't put a soldier out of action, much less a battle-armored Imperial Storm Trooper. To burn a one-square-centimeter hole right through a human body would require about 50,000 joules—optimistically assuming that all the laser energy would go into vaporizing tissue.

Once the size constraint is removed, it becomes possible to build a laser big enough to kill a person. The military has some lasers that are capable of the task; in fact, some surgical lasers are capable of slitting the throat of an immobile patient. To kill a soldier under realistic bat-

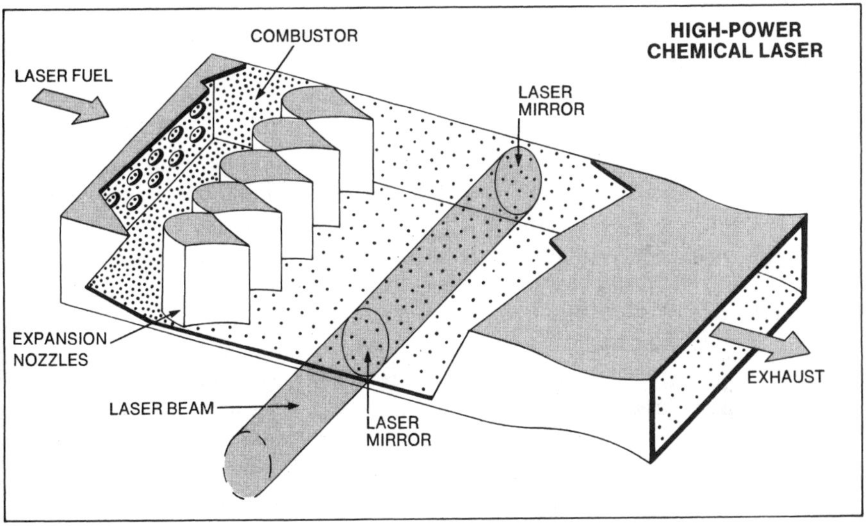

8. In a typical high-power chemical laser, fuels such as hydrogen and fluorine are prepared in a combustor and are then ignited to form excited hydrogen fluoride. The hydrogen fluoride passes between two laser mirrors, and the excitation energy is extracted to form a laser beam. To improve performance, atomic fluorine is produced in the combustor and is mixed with hydrogen at the expansion nozzles; the chemical reaction then occurs as the gas speeds to the right. The pressure drops after the gas passes through the expansion nozzles; this is important because the hydrogen fluoride laser works best at very low gas pressure. The low pressure makes it hard to exhaust the gas directly to the atmosphere, which is probably just as well, because hydrogen fluoride turns into corrosive hydrofluoric acid upon contact with water in the atmosphere.

tle conditions, though, you'd need a laser at least as big as a tank. As soldier-killing tools, guns have lasers beat hands down. Members of the National Rifle Association can rest easy.

REAL-WORLD LASER WEAPONS

Real laser weapons have a lot in common with wind tunnels. That's one reason why large aerospace companies are helping to develop them. Lockheed Missiles and Space Company, Rockwell International Corporation's Rocketdyne division, and TRW, Inc., are three examples. The laser medium used is a gas, which flows rapidly through specially designed nozzles. In many lasers the flow is supersonic—faster than the speed of sound. After the gas has passed through the nozzles, the energy is extracted in a laser beam. The two types of high-power gas lasers that are considered the best candidates for laser weaponry are electric-discharge lasers and chemical lasers.

The active ingredient in electric-discharge lasers is generally carbon dioxide, which is excited by a high-voltage discharge or a beam of electrons. These lasers emit invisible infrared light at 10.6 micrometers (0.01 mm). This wavelength is transmitted well by the atmosphere but is strongly reflected by many metals. A major problem is the bulk and weight of most electrical power supplies.

Chemical lasers generally get their power from the chemical reaction between atoms of hydrogen and fluorine (see diagram 8). It sounds like a simple process, but it isn't. First, pure fluorine and hydrogen are difficult to handle. Fluorine has some nasty habits, such as making stainless-steel tanks explode without warning. To prevent this from happening, the two fuels are generally put into more stable and easier-to-handle compounds. Second, the wavelength produced by a laser burning normal hydrogen is absorbed strongly by the atmosphere, so a special form—the rare heavy isotope deuterium—must be used, to produced wavelengths of 3.7 to 4.0 micrometers (0.0037 to 0.004 mm), which are transmitted better. Third, the reaction takes place at a pressure much lower than that of the surrounding atmosphere, making it difficult to get rid of the exhaust gas. That exhaust gas is a compound that itself would make a very nasty chemical weapon: hydrogen fluoride, which becomes hydrofluoric acid when it contacts water or even the normally moist atmosphere. Here we have a weapon that, if handled improperly, could kill not only the target but the users as well.

These are engineering problems the military is working on. The Army, for example, is trying to develop fuel and exhaust cannisters for chemical lasers. And the Pentagon is also looking further into the future at such potential contenders as excimer lasers, free-electron lasers, and even X-ray lasers, although each of these technologies clearly has a long way to go.

ATTRACTIONS OF LASER WEAPONS

These massive lasers obviously don't have people as their prime targets. To slip into government jargon, guns are more cost-effective antipersonnel weapons. However, war is changing from the simple slaughter of people. In modern warfare, it's becoming more important to kill hardware—enemy tanks, aircraft, ships, and missiles. Lasers are part of the trend toward the automated, electronic battlefield. They solve several major problems encountered with existing weapons.

First, when you fire a missile or a bullet at a moving target, you must obviously aim ahead of the target. To do that, you must accurately measure the speed of the target in all three dimensions, which is not always an easy task. Second, it's possible for the target to dodge your weapons by slowing down, speeding up, or changing direction. A related problem, familiar to anyone who's tried shooting ducks or clay pigeons, is that the force of gravity pulls a bullet toward the ground, making it necessary to aim above the target.

A laser beam travels at the speed of light—300,000 km (186,000 miles) per second. That's so much faster than any military hardware can move that the "leading" correction required is very small—typically negligible except over very long distances. "It takes six-millionths of a second for laser light to travel one mile, and in that time a supersonic airplane traveling at twice the speed of sound will travel only a little more than one-eighth inch," according to J. Richard Airey, director of directed-energy technology in the office of the under secretary of defense for research and engineering.

That's the most dramatic advantage of a laser weapon, but certainly not the only one. There are others:

- Because the beam is literally steered with mirrors, a laser can be shifted rapidly from target to target across a wide field of view. There's no need to pivot the entire device, as with conventional weapons—only the mirrors.
- Because a laser beam is straight and narrow, it can pick out and destroy a single enemy target among a group of many friendly vehicles.
- Each "shot" from a laser would require a relatively small amount of fuel (exactly how much is classified information). You could therefore store enough fuel for many shots.
- A laser beam can travel over long distances in space without losing energy and with little spreading. With proper lenses, it may be possible to destroy targets thousands of miles away.
- A laser could probably kill its target within a few seconds.
- Techniques designed to confuse the electronics in a guided missile

are not effective in foiling laser beams, though they could conceivably be used against the electronics that control the laser.

• Oddly enough, laser weapons may be relatively cheap. Although high-energy lasers are very expensive, the cost per shot is likely to be low, because a laser can shoot at many targets, whereas a missile can be shot at only one. Colonel Holmes, Airey's military deputy, explains that Patriot missiles cost $300,000 to $500,000 each, and even short-range Stinger missiles cost $20,000 each. A chemical laser burning the expensive heavy isotope of hydrogen, deuterium, would use up only $1,000 to $2,000 worth of fuel per shot, "getting into the realm of the cost of a burst of bullets." With an electrically-pumped carbon dioxide laser, the cost would be reduced to some hundreds of dollars per shot, according to Colonel Holmes.

TACTICAL WEAPONS

Laser weapons, like all military weapons, can be broken down into two types: tactical and strategic. Tactical weapons are designed to give advantage during battle. Strategic weapons are intended to give advantage before battle starts. In fact, one of the goals of developing strategic weapons is to convince the other side that it doesn't want to start a battle at all, because it will be sure to lose, or at least suffer severe casualties.

A space-based laser that could shoot down ballistic missiles shortly after launch is a strategic weapon. One of its purposes is to convince the enemy that it isn't worth trying to launch a missile attack. On the other hand, a tank-mounted laser designed to defend the tank and other targets would be a tactical weapon. We'll look at this type first.

Each of the three U.S. armed services has a program to develop tactical lasers. These programs have been around since about 1970, and all are aimed toward what the military calls "verification of lethality" by the early 1980s. In layman's language, that means laser developers must show that they can "kill" the types of targets the armed services want to kill and do so at a reasonable cost. It does not mean the weapons will be ready to be shipped to the battlefield. Rather, the demonstration must simply show that the laser will work as a weapon, at least in principle. Only then will the military start to design actual weapons.

The Air Force has the biggest program and is looking at laser weapons for bomber defense and tactical fighters. The showpiece of the Air

This winged drone was shot down . . .

. . . with a laser mounted in this modified Marine Corps tank. This Army laser is called the Mobile Test Unit. Courtesy Department of Defense

Force's program is the Airborne Laser Laboratory, a high-power carbon dioxide laser mounted in a military version of a Boeing 707 aircraft. "Mounted" isn't exactly the right word. The laser, built in the early 1970s, together with its power supply and auxiliary hardware occupies most of the cargo aircraft. (It's a gasdynamic laser, an older type of high-energy laser—in fact, the first breakthrough in powerful lasers. It has since been surpassed by chemical and electric-discharge lasers.) The laser shot down a winged drone (a remote-controlled target plane) in 1973 and was scheduled to shoot down an air-to-air missile aimed at it in airborne tests sometime in 1981. The Air Force didn't plan to take any chances with its expensive toy, however; the test was designed so that the missile couldn't really hit the Airborne Laser Laboratory.

The Navy's program, which was only half the size of the Air Force's program in 1981, is aimed at protecting ships against air attack, primarily by cruise missiles. In a test in March 1978 at San Juan Capistrano, California, the Navy destroyed an antitank missile with a chemical laser. Its next tests will be with a larger, deuterium fluoride, chemical laser called MIRACL, for Mid-Infrared Advanced Chemical Laser. MIRACL will attempt to protect a ship against an incoming cruise missile in simulated tests early this decade at the White Sands Missile Range in New Mexico, where the Department of Defense is building a new High-Energy Laser National Test Range. The beam-direction system alone, called Sea-Lite, will cost over $10 million. The laser will probably cost somewhat more.

The Army destroyed winged and helicopter drones with an electrically excited laser at its Redstone Arsenal in Alabama in 1976. That laser, called the Mobile Test Unit, was mounted in a modified Marine Corps LTVP-7 tank. The Army would like to use lasers on the battlefield against planes and airborne missiles and to defend "high-value" targets in the rear of the battlefield. That category could include lots of hardware—modern tanks cost over $1 million each. Military observers estimate the cost of a mobile battlefield laser at $5 million to $10 million. Actual development of laser weapons, of course, first requires that the Army sees the "convincing technical demonstrations" that Colonel Holmes says are a must.

STRATEGIC LASER WEAPONS

Laser weapons for strategic uses are fomenting the loudest controversy. Because strategic laser weapons could "kill" guided missiles and bombers, they could dramatically shift the balance of power, which is based on mutually assured destruction—the ability of the U.S. and the Soviet Union to obliterate one another in a nuclear war. High-energy laser systems, probably based in space, might offer a truly effective defense, not merely against ground-launched intercontinental ballistic missiles, but also against nuclear bombers and ballistic missiles launched from submarines. More on the controversy later; first, the weapons themselves.

The orbiting laser battle station is attracting the most attention. The first version to gain public attention surfaced in late 1979, when representatives of four defense contractors—Lockheed Aircraft Corporation, TRW, Inc., Perkin Elmer Corporation, and Charles Stark Draper Laboratory—convinced Wyoming Republican Senator Malcolm Wallop that such a system could be built "in the relatively near term." The "Lockheed Gang of Four," as the group was labeled by some critics, proposed putting 18 satellites equipped with high-energy lasers into orbit at an overall cost in the $10-billion range. Each satellite would have a 5-million-watt laser, a 4-m (13-ft) mirror to focus the energy onto a target, a pointer-tracker to direct the beam, a system to detect targets, and control electronics.

Wallop said that the "first-generation [system] could protect against all Soviet heavy missiles, about 300 other intercontinental ballistic missiles, nearly all submarine-launched ballistic missiles, and all long-range bombers and cruise-missile carriers." His estimate assumed that the Soviet Union would trigger all of its strategic systems around the world within 15 minutes.

There would be six laser satellites in each of three different orbits going over both of the earth's poles, providing essentially complete coverage of the earth. Each laser would be able to fire about 1,000 shots with a range "somewhat under 3,000 miles." The satellites would be armored, to protect them against nearby nuclear explosions, but would be vulnerable to attack from a high-power laser outside the satellite's range.

Recently, some further-out scenarios for satellites containing X-ray laser weapons have surfaced in *Aviation Week and Space Technology*. *Aviation Week*'s sources proposed satellites that would each contain an array of X-ray lasers and a nuclear bomb. In the event of enemy attack, each X-ray laser (a long, thin crystal) would be pointed at an enemy missile, and the bomb would be detonated to produce pulses from the lasers (in the process, destroying the satellite, of course). The idea was that the intense, ultrashort X-ray pulse would destroy missile electronics or perhaps even penetrate the shell of the missile itself. The proposals for weaponizing X-ray lasers generated strikingly uniform skepticism throughout the laser community; the politest comment was that these proposals were "premature." Government officials maintained a stony wall of silence.

The military is interested in laser satellites because of their obvious strategic importance. Moving lasers to space would also minimize the problem of getting laser beams to propagate through the atmosphere. However, DOD still considers putting high-energy lasers into space to be a "high-risk" concept because of the special requirements for launching anything into space: they must be light, compact, and able to function without human intervention.

Then there are those who think that laser weapons in space won't work in this century, and indeed might not ever work. Although conceding that it would be possible to build lasers with enough power to kill a target, Michael Callaham of Carnegie-Mellon University and Kosta Tsipis of the Massachusetts Institute of Technology maintain that it won't be possible to deliver the necessary energy to the target efficiently enough. In a report issued at the end of 1980, they also said it should be easy to develop countermeasures to make many targets invulnerable to laser attack, and that it would be extremely difficult and expensive to put laser weapons into space, because so much fuel would be needed. They also expressed concern that any effort to build laser weapons in space could trigger an attack by the other side—out of concern that such a system *might* work. The report recommended that the U.S. concentrate on developing countermeasures rather than laser weapon systems, and that only enough large lasers be built to test those countermeasures.

The debate is not an easy one to resolve. The people on both sides of the issue who tend to be the most vocal are usually those with the strongest biases clouding their judgments. Defense contractors obviously stand to make money from building new weapons. Conservative politicians, such as Senator Wallop, would like to have a dramatic new weapon to brandish at the Russians and are also inclined to push any military technology that the Russians are known to be pursuing. Military officials likewise tend to want more and better weapons, although the upper levels of the Pentagon bureaucracy have generally maintained a moderate, wait-and-see attitude toward laser weapons.

On the other side, political liberals often oppose expensive new weapons on ideological grounds, and scientists with liberal inclinations sometimes devote too much of their energy to finding reasons why new weapon systems won't work. There's also the phenomenon known as Clarke's First Law, after science-fiction writer Arthur C. Clarke, who wrote: "When a distinguished but elderly scientist states that something is possible, he is almost certainly right. When he states that something is impossible, he is very probably wrong." (For the purposes of discussion in his book *Profiles of the Future,* Clarke defined an "elderly" physicist as one over 30 years of age.)

Another possibility is suggested by Barry J. Smernoff, a Briarcliff Manor, New York, arms-control consultant: some high-up defense officials have become so used to the strategic status quo of mutually assured nuclear annihilation that they don't want to think that some technological development might make it obsolete.

DIPLOMACY, ARMS CONTROL, AND LASERS

At first glance, orbiting laser battle stations might seem to be a boon to global stability. They could destroy nuclear-armed missiles or bombers long before they could reach their intended targets. We would no longer have to live in a balance of terror.

The picture is not that rosy, however. While you and your family may not like the idea of all-out thermonuclear war with no defense, military strategists in the U.S. and the Soviet Union have learned to live with it. And it's their finger on the nuclear button, not yours.

Here's the problem. Let's say one side develops an antimissile laser system. That's fine, but it takes time to get it into orbit and to get it working properly. Callaham and Tsipis warn that if one side starts building a laser battle station, the other side would be strongly tempted to destroy it before it is completed, lest it find itself strategically outclassed. Destruction of the battle station could in turn trigger retaliation, perhaps all the way to an all-out nuclear war.

The Strategic Arms Limitation Treaty (SALT I) prohibits ballistic-missile defense systems, such as laser battle stations. But in the present political climate, there is legitimate concern that one or both sides might abrogate the SALT I treaty, much as they both ignore a 1967 treaty banning military uses of space, which they signed. Senator Wallop's proposed system, for example, would be flatly in violation of both treaties. However, existing treaties only outlaw *deployment* of antimissile systems, not their development.

Arms-control negotiators are well aware of the problems posed by laser weapons, not only for missile defense, but also for destroying satellites. Laser weapons were among the topics under consideration in the SALT II talks.

In an overview of the potential impact of laser weapons on defense strategy, Smernoff warned:

> Space-based laser weapons will *not* constitute a risk-free technological fix for the deep-rooted problem of nuclear weapons and nuclear war. What counts in the long run, however, is whether the United States can muster the degree of national wisdom and consensus it will need to meet the strategic challenge implied by the advent of such weapons and, at the same time, evolve a workable balance between military competition and political cooperation with the Soviet Union. Otherwise, future American leaders may wish that laser weaponry had been more of a strategic mirage than a revolution, so serious are its prospective implications for reshaping defense policy and arms control.

ANTISATELLITE LASERS

The first step toward using high-energy lasers in space would be weapons aimed at blinding or disabling satellites. Although not as critical as an antimissile system, an antisatellite weapon is important. Both the U.S. and Russia depend heavily on satellites to spy on each other. Satellites are used in military communication and navigation and in monitoring tests of nuclear weapons—and, of course, to warn of a launch of a ballistic-missile attack.

Satellites are relatively easy to kill. A high-power laser can blind a satellite's light-sensitive detector or damage the solar cells used to power some satellites. It may also be possible to disable satellites by destroying the sensor that keeps them locked in a stationary position relative to the surface of the earth.

There's no need to put these kinds of weapons in space: a high-energy laser sitting on the ground could be pointed at a satellite. Because the power required to blind a satellite is relatively low, the beam-direction problems that are one of the biggest obstacles to laser weapons would be minimized. Instead of having to focus on a spot only a few inches in diameter, the beam might be spread out over a much wider area, including the whole satellite.

Again, the biggest problem is figuring out when, or if, you've killed the satellite. That's not easy if the satellite continues to send signals, and if the signals are encoded so that you can't decipher them. One clue that the satellite is out of commission might be if the signal suddenly becomes uniform, because all the elements in the sensor are sending the same reading, meaning that it has burned out. However, it should be easy to design a satellite that encodes signals more complexly; and it's a good bet that the military shifted to that approach once they began considering the possibility of antisatellite laser weapons.

RUSSIAN WEAPONS

There have been reports, widely accepted in the defense community, that Soviet lasers have briefly illuminated U.S. spy satellites. That would not be surprising, but, if true, it does not mean that the Soviets have operational antisatellite lasers. It's one thing to hit a satellite once; it's another to do it repeatedly and on demand. Fifty Soviet tries at hitting a satellite might have gone by undetected before any succeeded. That kind of success isn't acceptable for critical military hardware. It has to work each and every time it's used. If, for example, the Russians wanted to shield a launch of intercontinental missiles by blinding our spy satellites, they'd need to *know* that the satellites were no longer operating—a far cry from just pointing a laser at the satellites.

But there's clearly a lot of work on high-energy lasers going on in the Soviet Union. Each issue of the monthly scientific journal *Kvantum Elektronika* (Russian for quantum electronics, the branch of physics covering laser operation) usually carries several papers on high-energy lasers or their effects on various materials. There are fewer such papers published in the unclassified literature in the West.

The exact size of the Russian high-energy laser effort is hard to estimate. Military sources put it at three to five times that of the United States, which spent about $200 million a year in 1978–80.

In testimony before a Senate subcommittee in late 1979, J. Richard Airey, the Pentagon's director of directed-energy development, said:

> The Soviet Union is apparently concentrating large resources on high-energy laser technology. In particular, they may be beginning the development of specific weapon systems. We, on the other hand, have decided to keep our high-energy laser program as a technology program for the next few years. We believe that we understand the technical issues basic to translating high-energy laser technology into weapon systems, that our decision is correct, and that the Soviets may be moving prematurely to weapon systems. However, we are continually conducting a careful review of our programs, as well as watching Soviet progress with great interest in a continuing reevaluation of this decision.

The difference in philosophy that Airey points out is a basic one. The Soviets sometimes pour massive resources into an effort with very specific objectives, while the U.S. tends to spend more time evaluating the possibilities (what Airey calls "technology programs"). Some critics of U.S. defense policy have charged that continued focusing on the technology—rather than turning promptly to making actual weapons—reflects a general lack of resolve. But it also reflects a rational conviction that there are great technical obstacles to be overcome *before* a laser weapon can be fielded by the military.

THERMAL BLOOMING . . . AND OTHER PROBLEMS

Building the laser is probably the easiest part of building a laser weapon system. Even critics Callaham and Tsipis have said it should be possible to build a laser capable of inflicting damage. The problem is getting the energy from the laser to the target.

The obvious problem is the atmosphere, which is nowhere near as efficient at transmitting light as it looks. Clouds, rain, snow, fog, and smoke can all obstruct laser beams. While the attacker can pick the weather best suited for his weapons, that luxury isn't available to the defense, and lasers are envisioned mainly as defensive weapons. For example, it might make sense to attack a laser-equipped ship on a foggy morning, when the laser beam might not get very far. It's also possible to lay down smoke screens on the battlefield that would hamper laser weapons; one example is fog-oil, produced by burning diesel fuel inefficiently. While it may be possible to burn through clouds, fog, and snow with a high-energy laser, weather remains high on the list of problems for tactical battlefield lasers.

Even on a clear day, the atmosphere absorbs a small fraction of the light passing through it. It's not noticeable with sunlight, but it can be a problem with a high-energy laser beam. The beam heats the air slightly, causing the warm air in the center of the beam to expand slightly and become less dense than the air around it. The result is that the air in the path of the laser beam acts as a very weak lens, which has the effect of spreading the beam out. Although the lens effect is weak, it becomes significant when the laser beam has to travel long distances and be focused sharply onto a small spot on a target. The phenomenon is called *thermal blooming* (because it's a heat-produced process that causes the beam to spread, or "bloom").

There are other atmospheric problems. The beam tends to wander off the target because of air currents and other random fluctuations in the atmosphere. Also, it's possible for the laser beam to be *too* powerful. Above a certain level, the air molecules tend to break down into free electrons and ions, forming a protective plasma around the target that absorbs the beam. Material evaporated from the target can do the same thing. All these processes interact to make atmospheric propagation and beam-target interactions complex subjects.

RUBBER MIRRORS AND FIRE CONTROL

Fortunately, there's a way around these problems. The basic solution is to shoot a laser beam—or some other kind of light beam—up at the target to see what the atmosphere does to the beam, then adjust for those distortions. The preferred approach uses mirrors.

In the vernacular of laser developers, these are called "rubber mirrors." They have flexible surfaces, which are precisely controlled by computers. Extreme precision is demanded, as the mirror's shape must be controlled within a tiny fraction of a wavelength of light and adjusted around a thousand times per second to compensate for the continual fluctuations in the atmosphere. The sensing equipment, in turn, must gauge precisely the conditions in the atmosphere, rapidly process this large volume of data, and continually adjust the ultraprecise mirror. While there has been a lot of progress, both the sensing/control equipment and the mirrors themselves need considerable work before they'll be ready for practical use.

The mirrors, by the way, must be huge. Laser-weapon designers talk blithely about mirrors 4 m in diameter (that's roughly 13 feet!). That's half as big again as the Mount Wilson 100-inch telescope mirror and almost as big (75 percent) as the monster 200-inch telescope mirror at Mount Palomar. Mirrors for both these telescopes are delicate, weigh many tons, and are rigid. A mirror for a laser weapon, on the other hand, must be flexible and light enough to make the whole system portable; it may even have to survive a launch into space without changing shape in ways that aren't intended. Such a mirror is beyond the state of the art in optical manufacturing. Yet 4 m in diameter is near the lower end of the scale for space-based laser weapons. Some developers are talking about mirrors as large as 30 m (nearly 100 ft) in diameter for use in space.

You also need what's called a *fire-control system,* to aim the beam at the target, fire the laser, and make certain the target has been killed. There's a lot more involved than what you see in the movies, where the hero simply lines up the target in a viewer, pushes the button, and waits to see if the target blows up. The fire-control system must find a vulnerable point on the target (and there may not be many), adjust the output mirror to compensate for atmospheric conditions, and focus the beam onto the vulnerable point—and keep it focused there long enough to destroy the target. (A series of laser pulses is worth little unless all the pulses are directed at the same area.) Then it must watch to see if the laser killed the target, which probably won't be obvious unless the fuel tank or something blows up. The result is a complex and expensive piece of hardware whose cost is comparable to that of the laser.

Taking the laser out of the atmosphere would simplify some problems in beam-steering and control. However, it's still no minor task to focus a beam with an accuracy of several centimeters (or inches) on a target over a thousand kilometers (several hundred miles) away and keep it focused on the same point. A space-based laser system for missile defense, for example, might have only 15 minutes to kill hundreds of enemy missiles, each carrying a deadly nuclear warhead.

This whole complex system, whether ground-based or space-based, must then be "militarized," so that it can be operated and maintained by ordinary soldiers, not people with Ph.D.s in laser physics. The laser has to be able to survive hostile conditions and operate reliably (that is, every time it's supposed to destroy a target). Finally, it must be affordable, or, in the military planner's terminology, cost-effective. It must provide the required amount of "bang" per military buck.

COUNTERMEASURES

While all this is going on, other military researchers are working on ways to foil laser weapons. They're trying to develop special coatings that would reflect enough of the beam so that little or no damage would be done to the target. They're also working to develop surfaces that would be easily vaporized to produce a smoke screen of sorts, to shield the target from the laser beam.

Callaham and Tsipis mentioned the possibility of launching a multitude of decoy targets along with a volley of missiles, as a way of confusing a laser weapon's fire-control system. Indeed, Callaham and Tsipis urged that the major thrust of U.S. laser weapons development be toward countermeasures rather than the weapons themselves, so impressed were they with the possibilities in this area.

Finally, the laser weapons themselves must be defended. If the Soviet Union, for example, were planning a missile attack, and the U.S. had laser battle stations in space, no doubt some of those missiles would be earmarked for our laser weapons, not just Omaha and Washington. We would also have to defend battle stations against other antisatellite weapons, which would include other lasers. Each battle station would need a full complement of laser countermeasures itself.

It's likely that the hardest job would be protecting a laser battle station in space during its construction. Barring dramatic breakthroughs in size and complexity, a laser battle station will not be launched into orbit ready to do its job. Rather, it will be put into space in pieces and assembled there. Such a construction project would be too big to hide from the enemy and would be unlikely to have effective defenses against all-out attack.

We should explain here that we don't mean to paint a pessimistic picture for the future of laser weapons, á la Callaham and Tsipis. We are only explaining the difficulties to be overcome. But it's pertinent to note that back in the 1940s many physicists had an equally long list of obstacles to building an atomic bomb.

A LASER BATTLE SCENARIO (SAY GOOD-BYE TO *STAR WARS*)

It should be obvious by now that the laser battle scenes in *Star Wars* and other science-fiction movies have nothing to do with what a real laser battle would be like.

For one thing, laser weapons will be *big,* and even tactical weapons will probably require a tank or airplane to haul them around the battlefield. There may be a soldier with a fire button, but more likely there will be an electronic system to home in on the target, once it has been identified either manually or electronically.

Either a rapid series of high-energy pulses or a continuous beam would be fired at the target. The laser would be "on" each target for a few seconds, then the fire-control system would verify the kill and turn the laser to another target, or try again. If the target were a missile, airplane, or spaceship, the beam would probably burn or break a hole through the skin, raising smoke from the surface. There probably wouldn't be wide misses. Instead, failures would come from an inability to keep the laser focused on the same spot and/or an inability to pinpoint a weak spot on the target.

Don't expect brilliant flashes of light through the air or space in the

Soldiers use a laser target designator to pinpoint a target for aircraft or laser-guided weapons. With this device, the soldiers can designate a target 6 km. (4 miles) away. Courtesy Hughes Aircraft Company

wonderful new world of laser warfare. The first generation of laser weapons will almost certainly emit infrared beams, which are invisible, and which will only manifest their presence by making targets glow or by burning up any dust particles in their way. Atmospheric scattering might make it possible to see a visible light beam close in to the earth, but not in outer space. Unless a space battle takes place in a particularly dusty corner of the galaxy, no one will see flashing laser beams—just flashes of light where visible beams are reflected off targets, or where a laser beam, visible or invisible, heats a target enough for it to glow. The only time a kill would be obvious would be when a fuel tank was ruptured and exploded; other hits might result in a spray of water or air if a tank was punctured, but these wouldn't necessarily be fatal.

If a laser weapon was designed to be controlled by a person rather than electronics—and humans tend to be necessary participants for drama—the apparatus would probably contain a high-power infrared laser weapon and a low-power visible-beam laser to aim the weapon. Once the operator, using the low-power laser, saw the spot at the desired place on the target, he'd pull the trigger (or push the fire button or whatever), and the high-power beam would follow the same path to the target.

All very exciting—but not as spectacular, or as noisy, as the movies, or even those "Space Invaders" video games.

ELECTRO-OPTICAL WARFARE

So far we've talked about lasers as weapons in and of themselves. But lasers have already revolutionized warfare in their role as accessories to conventional weapons. Here we're talking about low-energy lasers, which have become an important part of the military's arsenal of tools for the electronic battlefield.

Battle today is vastly different from that of twenty or thirty years ago. Soldiers no longer rely on the moon to see at night; they have infrared viewers that can "see" body heat and image intensifiers that can amplify low levels of light. Artillerymen no longer rely on triangulation and seat-of-the-pants techniques to aim their weapons; they have laser range finders to measure distances to targets and computerized hardware that controls how weapons are fired. Bombs are no longer simply dropped from airplanes in the hope that they will fall on the right target; they home in on a spot illuminated by a laser designator.

Much of this hardware is called electro-optical, because it picks up information in the form of infrared or visible light. A soldier may use electro-optical instruments to control a weapon, or the instruments may feed into a computer that controls the weapon. One of the military's long-term goals is "fire-and-forget" capability—being able to fire a weapon—say, a missile—and have it home in on its target without further human attention.

The laser was first used on a battlefield in 1972, with the introduction of the first laser-guided "smart" bombs in Vietnam. With these, a soldier on the ground (usually) illuminates the target with a low-power laser (see diagram 9). A sensor on the bomb detects the spot and directs the bomb as it falls, steering it toward the laser-designated target. The soldier doing the designation, as it's called, need not be close to the target—and wouldn't want to be, since a bomb is going to be dropped on it. All he needs is a clear view of the target. Designation can also be performed from the air.

The first lasers used as designators were ruby lasers, similar to the very first laser built by Theodore Maiman. While some ruby lasers remain in the military arsenal, they've been largely superceded by lasers built around another synthetic crystal, neodymium-YAG. The trouble with ruby lasers, besides the fact that they can't be pulsed very often, is that they have a nice red beam. When the enemy sees a red spot in his neighborhood, he has a pretty good idea of what's going to happen next. YAG lasers emit a near-infrared beam, invisible to the human eye but readily seen by electronic detectors. They can also be pulsed more rapidly than ruby lasers.

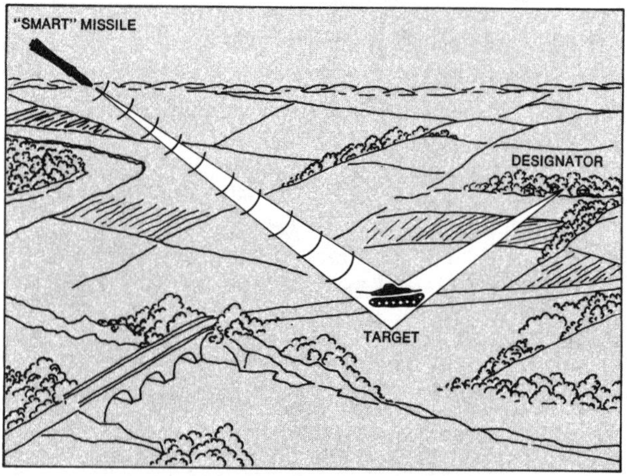

9. A soldier using a laser designator can hide in a safe place (in this case, among some trees on a hill) and point his laser beam at the target (in this case, a tank). A "smart missile" (or bomb) can then home in on the target.

The bombs themselves are generally ordinary (nonnuclear) bombs to which laser-seeking guidance "kits" are attached. The seeker is a simple electronic light detector, designed to "see" only the laser wavelength and, in its simplest form, divided into four quadrants. When the bomb is homing in on the laser-designated target, the same amount of light will fall on each of the quadrants. If it is off-target, one of the quadrants will receive a stronger signal than the others, and that signal will be used to correct the course.

The bolt-on kit is relatively inexpensive, and originally cost about $2,500, though the price has risen with both inflation and refinements to the hardware. The Air Force considers smart bombs highly effective, saying that one such bomb can do the job of 100 ordinary "dumb" bombs at only 10 percent of the cost. (If you look closely at those figures, however, you'll see that each smart bomb costs ten times as much as an ordinary bomb.)

The laser designator-seeker concept has been extended far beyond bombs. A whole new generation of artillery shells and short-range missiles that can home in on laser-designated targets has been developed. The Army has a cannon-launched projectile called Copperhead and a short-range missile known as Hellfire, which can be launched from a tank. The Navy also has a laser-guided projectile. Both the Navy and the Air Force have laser-guided bombs.

RANGE FINDERS AND FIRE CONTROL

Laser range finders determine the distance to a target by measuring the time it takes for a laser pulse to travel there and back again. A laser range finder can be a small hand-held unit only a little larger than a pair of binoculars, or it can be part of a large fire-control system for a tank.

Range finders are important even with laser-guided missiles and shells, because laser seekers can only provide fine adjustments. The projectile must be launched in the right direction, or it will never see the laser-designated spot on the target. Input from a range finder may go directly into the fire-control electronics or may be read out by a soldier and relayed to the soldier firing the weapon.

LASER RADAR

Lasers can also provide a kind of radar to give soldiers a picture of part of the battlefield. Carbon dioxide lasers emitting infrared beams might be able to identify enemy tanks and other targets.

A laser might also be able to identify an object as friend or foe. Such a system would identify the characteristic vibrations of a tank, say, as those of a friendly model or an enemy version. Another possibility is an active system in which a laser signal would trigger a response from a tank or other object; the proper response would identify the object as friendly.

Laser radar is now being used to measure concentrations of atmospheric pollutants for civilian applications; the military, on the other hand, is interested not only in ordinary pollutants but also in what one observer called "some very special pollutants." He was talking about chemical-warfare agents and special smokes that could be used as countermeasures against lasers and other electro-optical equipment.

COUNTERMEASURES AND COUNTER-COUNTERMEASURES

Just as there are countermeasures for high-energy lasers, there are countermeasures for low-energy lasers. The first level is simply the ability to detect that a target is being illuminated by a laser. Obviously, this warns that the target is about to be attacked.

A second level of countermeasure is detecting the source of the laser, which allows you to stage a preemptive attack. Both types of countermeasures have already been demonstrated but are still in development. Some military sources have wondered aloud if certain countermeasures might be more expensive than replacing destroyed hardware, but (not surprisingly) they did not mention the replacement cost of soldiers.

Naturally, the military is also looking at counter-countermeasures, ways to foil enemy countermeasures. One approach is a laser whose wavelength can be changed. This would help, because many proposed countermeasures rely on knowing what laser wavelength would be used, so that a sensor that is insensitive to sunlight but sensitive to that particular laser wavelength can be designed.

LASER-AIMED RIFLES

The military has done little work on laser techniques for aiming rifles. But the possibilities have not been lost on civilian police agencies and the companies that sell them equipment.

Several companies supply rifles with laser-aiming accessories or sell bolt-on laser-aiming units for older rifles. These are sold to such agencies as the FBI, local police departments, and other law-enforcement groups around the world.

The basic idea is simple. A helium-neon laser is aligned with the barrel of the rifle so that the red laser beam will travel the same path as the bullet. When the laser is turned on, its red spot pinpoints where a bullet would hit if the gun were fired. It isn't even necessary to sight down the barrel of the gun—just bring the laser spot to the desired point. That can be a big help to cops or agents who don't have the luxury of time to aim carefully. Some officials report that the bright red spot from the laser is also effective in convincing people resisting arrest to surrender, because they know that the police have them in their sights.

BATTLE SIMULATION

A variation on this device can also be used in war games. Up to now, keeping score in military war games has been a rather subjective matter, requiring human referees. To overcome this problem, the DOD is spending some $100 million on MILES, the Modular Integrated Laser Engagement System, capable of "arming" some 39,000 soldiers and about 6,000 tanks.

MILES is a simple but ingenious concept. Instead of firing bullets at each other, soldiers fire pulses of light. Each gun, including artillery, used in the war game is equipped with a small semiconductor laser that emits a coded series of low-energy pulses when the gun is "fired." The coding identifies the type of gun firing the shot, which is important, because some targets are supposed to be resistant to attack by certain weapons. For example, it wouldn't be realistic for a rifle to kill a tank, but an artillery shell or bullet could kill an individual soldier.

Sensors attached to each soldier and vehicle in the mock battle watch to see if they are hit. If the sensor detects the proper type of light pulse, it will tell a soldier he's been hit and disable his laser gun. In the case of a tank, the sensor tells the tank's crew that they've been killed, takes the tank out of the laser battle, and produces a plume of purple smoke to announce to everyone the tank's demise. The sensors are also interconnected to automatically keep score on the progress of the battle.

The U.S. system, made by Xerox Corporation's Electro-Optical Systems Division in Pasadena, California, isn't the only one. A British company has been making similar systems for several years and has shipped them to armies around the world.

Note that you still need referees in these battles, to make sure the soldiers don't make themselves invulnerable by covering their MILES sensors with Band-Aids.

A soldier holds an M-16 rifle that fires blanks and is equipped with a small pulsed laser, which he directs at his "enemies" during war games. Courtesy Sandia Laboratories

OTHER MILITARY LASERS

Lasers can also be used in military communications. One prospect is to use satellites and lasers to communicate with submerged submarines carrying strategic missiles.

Such subs are relatively safe as long as they stay underwater. But they are hard to communicate with when submerged. Plans to use very long radio waves to reach them have been impeded by environmental and political problems centering on the need for a large amount of land for a broadcasting antenna and uncertainties regarding the health hazards of such radiation.

As an alternative, the DOD is pushing development of lasers emitting beams in the blue-green region, which is the kind of light that seawater transmits best. One idea is to put the laser in a satellite and have it scan regions of the ocean where submarines are supposed to be. An alternative would be to leave the laser on the ground and shine the mirror up at a mirror on the satellite, which would then reflect the beam back to the ocean. The beam would not stay directed at a single submarine, however, because that would pinpoint the sub's location for the enemy. Even so, in theory such a system should be capable of transmitting enough information to the submarine. In practice the problem is the need to develop suitable lasers and detectors.

The military is also finding increasing uses for fiber optics. Optical fibers are small and light, critical factors in the design of a portable battlefield communications system. They are also immune to electromagnetic noise, which can jam other types of communication, and they don't emit any telltale radio-frequency signals. This means that the enemy can't eavesdrop on conversations; it also means that enemy radiation-seeking missiles can't home in on the communications base.

8 MAKING IT WITH A LASER: LASERS IN MANUFACTURING

When the laser was young, it was called "a solution looking for a problem." The solution has now found many problems in industry. In factories around the world, lasers are doing jobs ranging from drilling holes in diamonds to drilling holes in baby-bottle nipples. Lasers are cutting sheets of titanium for military aircraft and cloth for men's suits. Engineers have also discovered what lasers are *not* good for: jobs such as slicing fish or trimming lawns.

We should warn you in advance that this chapter focuses on only one facet of the use of lasers in industry—*materials working.* That's a deliberately vague term and includes any process that changes the dimensions or physical characteristics of an object. It's obviously an important part of industry, but it's not everything. Industry also requires measurement and inspection, and the ways in which lasers do those jobs are described in the next chapter.

The first boomlet in laser materials working came in the early 1960s. Richard Barber, an applications engineer with Photon Sources, Inc., a Livonia, Michigan, maker of industrial carbon dioxide lasers, summarized it in a few sentences:

> The first material processing that was ever done [with a laser] was with the embryonic ruby lasers in 1962. The technique was simple. Put a lens in front of the beam, focus it on a razor blade, and drill a hole. Indeed, those early lasers were frequently rated in Gillettes, an unofficial number indicating how many stacked blades could be pierced. Of course, it soon became popular knowledge that lasers could vaporize and drill holes in anything!

After *that* stampede passed and its cloud of dust settled, engineers and scientists of a more practical sort picked themselves up and began to ask and answer the questions: what size hole, how fast, through how much, what material, and what shape? The answers, of course, disappointed many people, and by 1966 the number of real processes for laser-drilling holes settled out at a handful.

The basic problem is one that's simple to define but hard to get around: what can be done in the laboratory often can't be done cost-effectively on the factory floor. The laser may not be reliable enough to be put on an assembly line, where it has to drill, cut, or weld each and every time a part passes by, all day along. The laser may simply be more expensive to use than an alternative technique. Or the quality of the hole (or weld or whatever) may not meet the required specifications.

However, the laser does have many important advantages that have motivated engineers to keep working on ways to use it. The laser is a noncontact machine tool: nothing touches the part being worked on, and there are no saw blades or drill bits to break or keep clean. This also makes the laser attractive for drilling tiny holes (because thin drill bits are weak) or cutting patterns too complex for a saw to follow. The absence of physical contact also makes lasers useful for machining materials that are hard and/or brittle, such as ceramics and even diamonds, or soft and easily deformable, such as rubber. The laser can deliver energy precisely to a tiny spot with little effect on the surrounding regions, and the beam can reach into otherwise inaccessible places—advantages also relevant to a very special type of materials working we described in chapter 5, surgery. The laser is also compatible with computer controls, which are growing increasingly important as factories continue to become more automated.

Lasers also have some serious disadvantages. As you might expect, one is price. Lasers with output power high enough to use in materials working are expensive. A modest, pulsed, hole-drilling laser costs well over $10,000, and a high-power laser with continuous output capable of welding or cutting thick sheets of metal can cost over half a million dollars. These prices put lasers into the big leagues of machine tools. As long as an engineer can get the job done with a less-expensive tool, the laser won't be put to work.

There are some things that lasers simply aren't very good at. One is drilling large holes. In general, the larger the hole, the more laser energy is required—and the higher the laser energy, the higher the price. The price of mechanical drills doesn't increase in the same way, so they become more attractive as the holes get bigger.

There's also a subtle psychological factor. Manufacturing engineers tend to be inherently conservative and stay with a process that works, as long as they see no clear advantages to other methods. That's only reasonable and rational, but it does tend to limit the use of lasers. In addition, lasers have sometimes been unable to survive the harsh environment of the factory floor, where dust can contaminate optics, and heavy vibrations can disturb optical alignment.

At this writing, we estimate that somewhere around 10,000 to 20,000 lasers are being used for materials processing around the world, with most of them in developed countries, such as the U.S. and Japan. Sales of materials-working lasers around the world (other than in communist countries) were estimated at $70 million in 1980 by *Laser Focus* magazine. That translates into somewhere in the range of 2,000 to 3,000 lasers sold that year and represents a growth rate (in dollars) of about 25 percent a year. Although that may sound like a lot of lasers, they're not distributed widely. Instead, they're concentrated in large factories or in companies called "job shops," which specialize in performing difficult materials-working tasks for other companies.

WHAT DO LASERS DO?

The jobs done by lasers can be categorized by the amount of energy needed. Cutting and drilling are the most demanding tasks, as the laser must deposit enough energy, albeit on a small spot, to vaporize the material. Lower powers are required to write a message or code a number or trademark on something, for example, to write a serial number on an automotive part, because only a surface layer has to be vaporized.

When the amount of laser energy deposited is even lower, the material melts rather than vaporizes. In this case, the laser becomes a welder, used to join two pieces of metal together. Lower powers still are used for heat-treating or *annealing,* where the internal structure of a material is changed by heating without melting. Controlled heating of certain steels, for example, can toughen their surfaces to help them withstand wear.

A high-energy laser blasts through an industrial diamond. Courtesy Hughes Aircraft Company

Both pulsed and continuous-output lasers can be used for materials working. Pulsed lasers are generally used for drilling because of the need for short, intense bursts of energy. Welding and cutting can be done with either pulsed or continuous-output lasers. Heat-treating is generally done with continuous-output lasers.

Most materials working requires high powers, and only four types of commercial lasers are powerful enough: neodymium-YAG, neodymium-glass, ruby, and carbon dioxide. Of these, YAG and carbon dioxide can be made to operate in pulses or continuously, whereas ruby and glass lasers can work only in pulses. Carbon dioxide lasers can produce the highest powers in continuous beams.

DRILLING HOLES

Lasers can drill holes in diamonds, and it's done all the time. But it's not a stunt to show off the laser's prowess. Because diamond is the hardest material known to man, it is used in drawing wires. Wires are smoothed and shaped by punching a hole through a diamond and then forcing, or drawing, the metal through the hole.

Before the advent of the laser, drilling a single hole through a diamond could take up to two days. Today, a technician fires a series of laser pulses and evaporates a hole in the gem in a few minutes.

One problem is that the laser does not leave the sides of the hole smooth enough for drawing wire, and some mechanical polishing is required. Nonetheless, the whole process is still much quicker than it was before.

Ceramics can also be drilled with lasers. Ceramics are finding many uses in electronics and other industries, because they're rigid, lightweight, strong, can withstand high temperatures, and don't conduct electricity. However, drilling holes in, or attempting to cut, ceramics is difficult. Ceramics are brittle as well as rigid, and a drill can easily break them. Ceramics can also wear drills down rapidly. A further complication is that it's nearly impossible to bore the small holes required in microelectronic circuits with ordinary mechanical drills. Lasers can drill holes in ceramic circuit boards that are as small as one twenty-fifth of the thickness of the material. In contrast, mechanical drills often break when the material being drilled is thicker than the diameter of the hole.

Lasers are particularly attractive for drilling circuits boards, because they can be programmed to bore a series of holes automatically. All human operators need do is load material into the laser system; once the laser has been told where to drill the holes, it doesn't need further human guidance.

Drilling a series of holes in a ceramic with a laser can serve another purpose—perforating it so that it can be neatly broken apart. In this case, the holes penetrate only partway through the ceramic and are drilled in a straight line. When the ceramic is bent, it usually breaks along this "dotted line." Such laser scribing is often the easiest way to cut brittle ceramics into pieces.

HOLES IN SOFT MATERIALS

Rubber and plastics are at the other end of the hardness spectrum from ceramics and diamond, but they can also be drilled with lasers. Ordinary punching or molding techniques work well for big holes but not for tiny ones.

Carbon dioxide lasers have been drilling holes in baby-bottle nipples since the mid-1960s. It takes only about 15 watts for the laser to punch a hole through the rubber, and the hardest task in designing a laser system for this purpose is building equipment to pass the nipples through the system so that they can be drilled rapidly. Production rates of about one nipple per second are achieved readily, and the holes are much more uniform than holes produced mechanically.

If you don't have a baby around the house, you probably still have tiny laser-drilled holes—in aerosol valves. Relatively large holes were used in valves for aerosol cans with fluorocarbon propellants. But when people started worrying about possible depletion of the atmosphere's ozone layer, there was a shift away from fluorocarbons, and engineers had a problem. Alternative propellants required valves with smaller holes—only 0.15 to 1 mm (0.006 to 0.040 in) in diameter.

The smallest holes were very difficult to make in the conventional manner, in which the plastic is molded into shape with metal pins forming the holes. The smaller the hole, the more likely the pin is to break. So engineers turned to lasers. Carbon dioxide laser beams of 50 to 185 watts can produce such holes in 0.001 to 0.050 sec. With two parallel production lines running past the laser, it can produce up to 600 to 700 valves per minute. Thus it becomes possible to chalk up another, albeit dubious, achievement for the laser—helping to save the aerosol spray can.

The holes in baby-bottle nipples and aerosol cans are small, but there are even tinier laser-drilled holes you may never notice, even though they're in common products. For example, lasers punch tiny

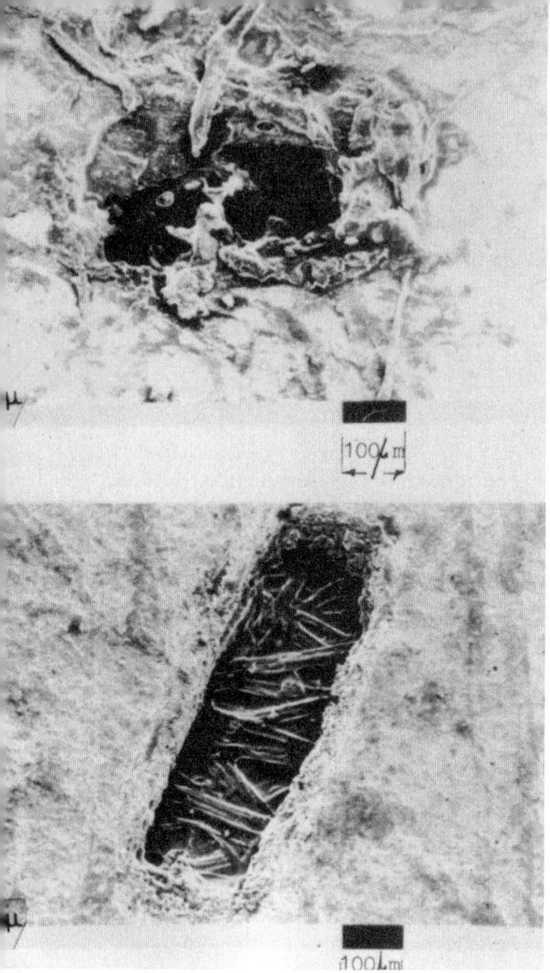

A comparison of holes cut in cigarette paper by a mechanical punch (*above*) and a laser (*below*) shows that laser drilling can be much more precise. Courtesy Coherent, Inc.

holes in cigarette paper. The tar content of cigarette smoke, which is prominently listed on every pack, depends on the air flow, and the precision of laser-drilled holes lets cigarette makers put a tighter specification on the amount of tar. For example, with ordinary perforation techniques, the amount of tar in one brand of cigarette might vary between 3 and 9 milligrams (mg), but with laser perforation, the range might be much smaller—say 5 to 7 mg. That would let the manufacturer list the tar content as 7 mg, rather than the 9 mg that would be required otherwise (government rules specify that the highest possible content be listed).

You'd need a microscope or a powerful magnifying glass to see these holes. They average 0.25 mm (0.01 in) in diameter and are 1 mm (0.04 in) apart. They're punched out by a carbon dioxide laser at rates approaching three million per second, as the paper is rolled past a line across which the laser beam scans, punching out holes. Coherent, Inc., of Palo Alto, California, says that every major supplier of cigarette papers has at least one of its laser systems, and that some have as many as fourteen.

Lasers are also the cleanest and fastest way to drill holes in drug capsules to permit timed release of medication, at least in the view of the Alza Corporation of Palo Alto, California. Felix Theeuwes, vice-president for product research and development, says that a single hole, 0.12 to 0.33 mm (0.005 to 0.015 in) in diameter, is drilled into each capsule. The actual size depends on the type of drug and how fast it is to be released into the body.

The laser can drill holes of precise sizes, which is critical for drug capsules, since the medicine will flow out too quickly if the hole is too large, and water seeping through the capsule's semipermeable skin will be retained and eventually burst the capsule, if the hole is too small. A system from Coherent, Inc., capable of drilling holes in 1,000 capsules per minute was being tested in 1980; in 1981 the company hoped to begin producing holes in millions of capsules per day with several such laser systems. The laser-drilled capsules will contain such common medications as anti-inflammatory and cardiovascular drugs.

DRAWBACKS TO LASER DRILLING

Although laser drilling can solve many problems in materials working, it is not without its limitations. One is the high reflectivity of metals, up to 90 percent or more, which pushes up the power requirements. To get a 90-percent reflective surface to absorb 50 watts, you need a 500-watt beam; if the surface reflects only 10 percent of the incident laser light, however, you need only a laser that produces a 56-watt beam—which is much less expensive and is easier to operate.

The material that the laser vaporizes creates problems, because it does not simply vanish. Most of it is generally ejected from the hole, but some of it condenses immediately around the outside of the hole, creating a crater-like ring. The focusing lens has to be protected from

A multibeamed carbon-dioxide laser can drill or cut through metal parts. Courtesy Coherent, Inc.

the ejected material. Also, in many materials, the hole itself isn't smooth, because the thin layer of material melted, but not vaporized, by the laser doesn't resolidify evenly, thereby causing cracks and weakening the material.

In general, lasers don't drill holes with perfectly vertical sides (see diagram 10). The problem is the way the beam must be focused—it comes to a point slightly below the surface of the material. In other words, the beam is cone-shaped and tends to create cone-shaped holes. This taper can be minimized in several ways, such as by picking a lens with longer *focal length,* so that the taper of the cone is more gradual, but it still presents a problem for many applications. As a result, many laser-drilled holes must be smoothed out, both to remove resolidified material and to make the sides of the hole more vertical.

The fact that the laser beam is tapered to a point also limits the ratio of the hole's length to its diameter. Lasers can drill holes as long as twenty-five times their diameter in ceramics, but in most metals a laser-drilled hole can be no longer than about ten times its diameter. Problems in getting the vaporized material out of a hole also set limits on its depth.

CUTTING WITH A LASER

A laser cuts nonmetals the way it drills them, by vaporizing the material. Usually there is a jet of gas blowing on the region being cut, its purpose being simply to get rid of the vaporized material by blowing it out of the way before it interferes with the cutting process or settles on the laser system's optics, thereby damaging them. The gas is generally air or a chemically nonreactive gas, such as argon—the latter is used when there is a possibility of a fire or an explosion.

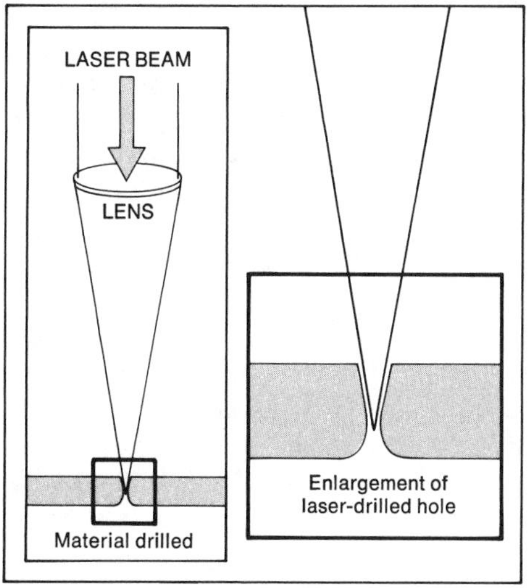

10. A drawback to laser-drilled holes is that they are conical rather than cylindrical. This is because the lens that focuses the laser beam makes the beam itself conical.

In theory, the same process could be used for metals. However, metals reflect laser radiation strongly, so attempting to cut them simply by laser vaporization would require prohibitively high powers. There's a simple, yet elegant, way to avoid that problem. It relies on a jet of gas directed at the same spot as the laser beam. But the gas is not pure air or inert gas—it's pure oxygen.

With the oxygen flow, the role of the laser is not to vaporize but merely to heat the metal to a temperature at which it will react with the oxygen. The reaction of the hot metal with oxygen burns away the metal to form a cut, and the gas jet carries away the debris. Thus, the process isn't, strictly speaking, laser cutting, but rather, *laser-assisted* cutting, with the actual cutting being done by the oxygen burning away the laser-heated metal.

Whatever you call it, laser-assisted cutting of metals is a dramatic process, which causes an impressive number of sparks to fly. It can also be very useful for cutting certain metals that are difficult to cut by conventional technologies. As a result, it's captured the attention of the military, for cutting the sheet metal used for aircraft skins, among other things.

Carbon dioxide lasers have come to be accepted as the best tool for cutting one problem metal—titanium. Tough, yet light in weight, titanium is used in some military aircraft. It's so tough, in fact, that it chews up blades that are used to cut it. It's also expensive, which puts a premium on cutting it efficiently.

Fortunately, titanium turns out to be easy to cut with a carbon dioxide laser. It has a low ignition temperature and reacts readily with the oxygen jet—permitting fast cutting. For example, a 375-watt carbon dioxide laser can cut 6-mm-thick (0.25 in) sheets of titanium at 250 cm/min (100 in/min). The laser-assisted oxygen jet cuts a much narrower swath than an ordinary saw would, reducing the amount of valuable titanium that's turned into "sawdust."

Lasers have also been called upon to solve some unusual metal-cutting problems. At the United Technologies Research Center, for example, brittle sheets of metal from the U.S.S. *Monitor,* a Civil War ironclad ship, were cut with a laser, because any other method would have shattered them. The cutting was part of a study of the feasibility of recovering the ship, which was sunk off Cape Hatteras in a storm over a century ago.

CUTTING PLYWOOD

Since the early 1960s, people have been intrigued by the idea of cutting wood with lasers, presumably in the hope that lasers would cut quickly and cleanly. Most of those hopes have been disappointed. Laser cutting almost inevitably loses to the traditional competition—saws and knives—because of its much higher cost, limited cutting depth, and the fact that it causes noticeable charring of the wood at the edges of the cut.

There's one notable exception: cutting of the plywood dies used to guide the knives that cut and crease cardboard for cartons. The standard "dieboard" that's used for this purpose is ¾-inch-thick maple plywood. The requirements are stringent—the slots for the blades have to be narrow and vertical, and the pattern cut by the blades has to match the pattern printed on the cardboard.

Before the laser came along, the job was typically done by hand. It would take a craftsman a week to cut a typical complex dieboard. With a 375-watt carbon dioxide laser, the same job can be done in a day. The cuts are computer-controlled with an accuracy of 0.4 mm per meter of cut (0.005 in per foot), and once the cutting pattern has been programmed, the laser system can repeat it on demand to produce a duplicate dieboard.

CUTTING RUBBER AND PLASTIC

Rubber, of course, is flexible, and this can make it difficult to cut; it bends under a knife and can be deformed. Where dimensions and shape are critical—as, for example, making a rubber seal for a vacuum or high-pressure system or for liquid-flow equipment—manufacturers sometimes turn to lasers.

A rubber seal may require a tilted, rather than a vertical, edge. Cutting such a part with a knife would be extremely difficult, because pressure from the blade would distort it. With a laser, however, all you have to do is direct the beam at an angle. There is no distortion, because there is nothing touching the part. And because rubber absorbs the light from a carbon dioxide laser efficiently, laser cutting is fast.

Lasers can also cut plastics. They're particularly useful for cutting plastic fabrics, because the heat from the laser beam can both cut the material and seal it at the same time. Automobile seat belts are one example of how fast laser cutting not only produces material of the right size but also seals it to prevent unraveling.

CUTTING CLOTH

A laser will never replace a sharp pair of scissors in the hand of a tailor or seamstress. But lasers are getting into the clothes business simply because they fit nicely into computerized systems for cutting patterns from bolts of cloth.

The key to the laser cloth-cutting system developed by the Hughes Aircraft Company's Industrial Products Division in Carlsbad, California, is computer control. Patterns are fed into the computer, which determines how to lay them out on the cloth so as to use as much of the fabric as possible. Of course, instead of actually laying out pieces of a paper pattern, the system moves the beam from a carbon dioxide laser around on a table where the cloth lies. The beam cuts out the pattern, as air is sucked down through the cloth to hold it in place and remove the smoke caused by the laser cutting. The process is clean and causes negligible charring of the fabric, which is usually used in men's suits.

DYNAMIC BALANCING

If you own a car, no doubt you've had your wheels balanced so that they turn smoothly without wobbling. The various wheels in certain types of machinery must also be balanced. The problem is that unbalance is best detected when a wheel or gear is rotating, whereas balancing requires removing material from the wheel, which ordinary machine tools can only do when the wheel or gear is stationary. Then you have to rotate the wheel rapidly again to check the balance and repeat this process over and over again to get it just right.

But a neodymium-YAG laser can remove the unbalancing material while the wheel or gear is spinning. The part is rotated at high speed so that the heavy region can be located, then the laser is triggered to vaporize a small amount of material. The process continues until the wheel or gear is balanced and spinning smoothly. At present, laser balancing can be done only on small parts, because the laser can remove only a little material at a time.

MARKING AND ETCHING

Lasers are also being used to write codes, serial numbers, and trademarks, and to draw pictures on various products.

The traditional means of marking code or serial numbers on industrial parts is to stamp the numbers into the surface with some sort of die. This is a cumbersome process at best. What's more, some materials are simply too hard—either the numbers don't sink in, or the die or the material being marked is liable to be damaged. Such stamping can also be difficult to automate, which is a serious problem as industry tries to increase automation on its production lines, and as government agencies establish rules requiring manufacturers to keep better track of their products by more elaborate coding techniques.

A laser can mark hard or soft surfaces comfortably. One approach is to write with a series of short, intense laser pulses. Each pulse removes a bit of material, leaving a dot on the surface. The part being marked is then moved slightly, and the laser is fired again, leaving another dot. The result is a series of dots that spell out a code or serial number. The laser systems that do this can be programmed to write a different serial number on each part passing by. They also perform the task quickly and can write letters in almost any desired size. Semiconductor manufacturers use lasers to write tiny numbers on wafers of silicon, which will be cut up into integrated-circuit chips for a vast variety of uses in electronics. The automobile industry is using lasers to write serial numbers on hard metal parts—sometimes in inaccessible places—in part to aid police in identifying stolen cars.

There's also another type of laser marking, in which the image is written not as a series of dots but as a continuous engraving. In some cases the process is analagous to engraving, with a laser beam scanning an object and removing material from the surface—say, from a piece of wood—as it goes. But in other cases the whole image is etched at once. In the latter approach, developed by Lumonics Research, Inc., of Kanata, Ontario, a pulsed laser beam is passed through a mask carrying the desired image, then the beam is focused onto the object to be marked. The focusing lens recreates the image of the mask on the surface, and the laser energy etches away material to form the image. The process is very similar to stenciling and is used for similar purposes in many applications where ordinary stenciling wouldn't work—such as on a bare rubber surface that can't be painted, or on tiny electronic parts.

LASER WIRE-STRIPPING

Lasers have also been used to strip insulation from coaxial cable, in which a central copper wire is surrounded by successive layers of electrical insulation, metal sheathing, and protective plastic material. Ordinary mechanical strippers are likely to nick the wire-mesh or foil sheath around the cable. There is no such problem with a carbon dioxide laser, because its beam is strongly absorbed by the plastic and insulation yet strongly reflected by the metal. This reflectivity automatically stops laser stripping far more effectively than the strength of the metal stops mechanical wire-cutters. However, the lasers aren't portable enough for a technician working in the field, so lasers are used for repairing defective cable only in cable factories.

There is one notable exception, though: a hand-held laser stripper was used in building the Space Shuttle. Although many times more expensive than ordinary cutters, the laser stripper was chosen because reliability was paramount, and it provided essential insurance against damage to metal wires during mechanical stripping of the special, tough insulation used.

LASER WELDING

All the jobs we've described so far require that the laser vaporize material. Welding requires less energy, because two pieces of material, generally metal, must only be melted and held together until the material solidifies to form a joint. Vaporization isn't merely unnecessary; it's downright harmful if it occurs within the weld, because it can create bubbles, which can cause the weld to crack and fail.

The laser has advantages over conventional welding with a flame or an electric arc. Laser welding doesn't require a special atmosphere; it can be done in ordinary air, in a vacuum, or in many special mixtures of gases. Because there is no physical contact, a laser beam can even pass through glass and make welds inside a sealed vacuum tube, without damaging the glass housing.

Welding can be performed with pulsed or continuous-wave lasers, either ruby, neodymium-glass, neodymium-YAG, or carbon dioxide types. With pulsed lasers the pulses are generally longer than those used for drilling or cutting. This is because the energy must be spread out—both over the surface and in time—so that material within the weld isn't inadvertently vaporized, thereby weakening the finished weld. Some vaporization of surface material is unavoidable; for that reason a jet of inert gas is often directed toward the surface to blow the vapor away from the region of the weld. Lasers that emit a continuous beam—generally carbon dioxide types—are a natural for laser welding, particularly of thick metals.

WELDING THICK METALS

The thickest metals that have been welded—at least in the tests that have been publicly reported—have been 2-inch-thick (50-mm) sheets of stainless and another type of steel. Those welds were made at a rate of 50 in/min (1.3 m/min) with a 77-kilowatt beam from a gasdynamic carbon dioxide laser at a facility operated by the Naval Research Laboratory (NRL).

That laser isn't exactly standard industrial equipment. It was made for military tests of laser weapon effects and can emit up to 100 kilowatts for a few seconds. (The highest-power commercial lasers with continuous beams made for industrial use are also carbon dioxide types, but their output is no more than about 15 or 20 kilowatts.) "No one will use such a system for industrial welding," according to Conrad Banas of the United Technologies Research Center in East Hartford, Connecticut, who conducted the 1976 tests. However, the tests do point the way to development of high-power industrial lasers.

Welding goes faster with thinner material. Speeds of more than 2 in/sec (50 mm/sec) were achieved when a 90-kilowatt beam from the NRL laser was used to weld 1.5-inch (40-mm) steel. The welds didn't meet the Navy's stringent tests for the quality of welds on a ship, but they weren't expected to, because they were the first welds made at such high laser powers. The Navy is interested in welding heavy sheets of metal for use in the hulls of large ships and submarines.

The Navy was encouraged enough by the results to plan more tests. A carbon dioxide laser capable of producing up to 25 kilowatts of continuous output is being installed in a military equipment production plant in Minneapolis, operated for the Navy by the FMC Corporation. Operational goals include welding 1.5-inch-thick (40 mm) pieces of mild steel and ¾-inch-thick (19 mm) sheets of carbon steel at rates of more than 30 in/min (760 mm/min).

Welding with high-power carbon dioxide lasers (generally defined as those with continuous output of 2 kilowatts or more) is also being studied elsewhere. Avco Everett Research Laboratory, Inc., in Everett, Massachusetts, and its metalworking lasers subsidiary in Somerville, Massachusetts, have built several large lasers with outputs in the 10- to 15-kilowatt range and now offer to build lasers designed for round-the-clock use on production lines. Those lasers are being used for research at places like the General Motors Technical Center, the IIT Research Institute in Chicago, the Italian automaker Fiat, and a machining research center in Italy. The Ford Motor Company has tested a multikilowatt carbon dioxide laser from United Technologies for welding automotive underbodies. Probably the widest range of applications is being studied at the IIT Research Institute, which performs research under contracts from a number of companies and government agencies. A major emphasis has been on welding copper, a metal difficult to weld by other means.

The ability to weld a copper-nickel alloy might pay some big dividends to the owners of merchant ships, according to a study performed by the United Technologies Research Center for the New York–based International Copper Research Association. Because the copper-nickel alloy is much smoother than the steel used for hulls in ordinary ships, making hulls of the copper-nickel alloy would greatly reduce the drag on a ship. It's also a material that barnacles (which add to both the drag and the weight) don't attach themselves to. According to L. McDonald Schetky, technical director for metallurgical research at the International Copper Research Association, spending $3.5 million to clad a 200-m-long (700 ft) ship with a copper-nickel alloy could save at least $60 million over the ship's 20-year lifetime in reduced fuel consumption, lower maintenance costs, and the increased number of trips per year that would be possible because the ship could go faster.

Despite all this interest in multikilowatt laser welding, there are only two places where such multikilowatt lasers are operating on production lines: at Gould Inc.'s Industrial Battery Division in Fort Smith, Arkansas, and at C & D Batteries, a division of the Eltra Corporation in Leola, Pennsylvania. In both places they're doing the same task—welding lead-acid batteries that the companies supply to the Western Electric Company, the manufacturing arm of the Bell Telephone System.

The lead-acid battery was designed by Western Electric to close tolerances, which helps make laser welding possible. Successful laser welding requires that the joint be precisely and tightly formed, because the laser welding process can't bridge a large gap between two metal sheets. In their first three years of operation, the lasers used in the battery project produced a total of 10 million individual welds in more than 100,000 batteries. Although that number may sound impressive, it represents a relatively small production in the battery business, but manufacturing engineers say that the small production is what helps make laser welding feasible.

LOWER-POWER WELDING

High-power welding is dramatic, but high powers aren't needed for attaching relatively thin sheets of metal to each other, and it's such low-power welding that has proved most successful.

Conventional welding is a complex process, requiring a skilled welder. In contrast, laser welding is simple. "The same optical train, gas nozzle, and shield gas will handle most metals, from carbon steel to alloys of zirconium, [and] there are no electrode materials, fluxes, welding rods, torch tips, polarities, etc., to be matched to each individual job requirement," according to Simon L. Engel, president of HDE Systems, Inc., in Sunnyvale, California, a company that specializes in laser materials processing for other companies.

As a result, ruby, neodymium-glass, neodymium-YAG, and carbon dioxide lasers with continuous or average pulsed output powers of 1 kilowatt or less can be found on many production lines. The list of small parts that have been successfully welded with such lasers includes batteries for heart pacemakers (where integrity of the seal is critical), orthodontic braces, faulty electrical connections inside sealed glass vacuum tubes, and even nickel electrodes welded to the body of spark-plugs.

LASER SOLDERING IRONS

Lasers can even be used for soldering. They won't replace the soldering iron in the home workshop—even of the most ardent electronic hobbyist. However, they can solve the manufacturing problems that arise as computers continue to shrink in size.

Many connections must be made among the multitude of tiny electronic components in modern computers. This becomes a serious problem as computers become smaller, and the packing becomes so tight that manual soldering techniques are difficult for anyone bigger than an elf.

Apollo Lasers, Inc. of Los Angeles has sold more than 35 laser soldering systems—altogether worth almost $1 million—in a single year. The systems use a small pulsed carbon dioxide laser to solder as many as 40 joints per second on printed circuit boards with as many as 100 connections per square inch of surface area.

Laser soldering, according to Apollo president Fred Burns, is up to ten times faster than manual soldering. Moreover, it produces better-quality joints, because it can apply the same amount of energy (typically one or two joules) to every joint—a repeatability that even the most skilled worker can't match with a soldering iron. If the wire is insulated with an enamel coating (rather than a plastic jacket), the laser can vaporize the insulation at the same time that it solders the joint, with a single laser pulse.

HEAT-TREATING

To understand the process of heat-treating, or "transformation hardening," metals requires a brief explanation of the structure of solids. Metals, like many other materials, can exist in a variety of solid states, each of which has a different internal structure. The principal differences between the states lie in the arrangement of the atoms within the solid. That state in which such a solid exists depends on the temperature and pressure at which it was formed and on what has happened to it since its formation.

The most familiar example is carbon. Under normal conditions, carbon forms graphite—a soft, black material with little strength. However, under very high pressures, graphite is transformed into diamond, the hardest crystal known to man.

Engineers often try to transform an *entire* metal component into a stronger—or otherwise more desirable—state. But this isn't always possible, and often it isn't necessary or beneficial. For example, some forms of steel have a tough surface but are so brittle that they would break easily if entire parts were made of them. Thus engineers may seek to harden part or all of the surface of a piece of metal but not its interior.

There are two ways to do this—by increasing the temperature or by applying pressure. Centuries ago blacksmiths combined the two by hammering a piece of hot metal to increase its strength. Today, continuous-output carbon dioxide lasers do the same thing with heat alone, on production lines at General Motors and elsewhere. And researchers are studying ways to strengthen aluminum alloys using the shock waves caused by laser pulses striking metal surfaces.

In industry, the metal is coated with material that absorbs the laser radiation much more strongly than the metal itself, since metals are generally highly reflective at the carbon dioxide laser wavelength. Then a laser beam is scanned across the surface. How much the metal is heated depends on how intense the laser beam is, how fast it is scanned across the surface, and how much of the laser beam is absorbed by the coating. The amount of heating, in turn, determines how thick a layer of the metal is transformed.

The laser approach has several advantages. It's energy-efficient (despite the relatively low efficiency of the laser itself), because the heat goes only where the beam is directed. It's less likely than conventional heating to warp the part being treated. It's self-quenching—that is, the metal cools down once the laser beam moves on. And the laser beam can be directed at hard-to-reach areas of the component.

Heat-treating has become a major application of carbon dioxide lasers emitting continuous beams of more than a kilowatt. Laser heat-treating has found a home in the automobile industry. General Motors alone has over twenty heat-treating lasers, which operate on components ranging from cylinder liners for locomotives to automobile camshafts.

WHAT DOESN'T WORK

Lasers may be the greatest thing since sliced bread for many industrial applications. But they aren't very good for slicing bread—or for a variety of other things. A lot of people have learned this the hard way in the years since the laser was invented.

The laser is a loser for cutting food, a laser engineer confided to one of us over lunch one day. The cutting part works fine—the problem is that no one wants to eat the food afterward. Engineers at his company had tried cutting bread—and had gotten burnt toast. They'd tried slicing raw fish—and had ended up with cooked fish. They'd tried drilling holes for lollipop sticks—and ended up with ugly charred holes in the candy.

There's a footnote to the burnt toast story. Later, over another lunch, another laser engineer delighted in telling us how his company had gotten rid of a baker who had arrived at its offices with several loaves of bread he wanted cut with a laser. Company engineers explained to him that they didn't do such things . . . but that they knew of a nearby firm that did. It was, of course, another laser company—an arch-rival. They told the baker that this particular firm had a specialist in the laser cutting of bread, then gave him the name of the company's president. We don't know exactly what happened next, but the rival company was the one that got the burnt toast mentioned in the preceding paragraph.

Some other applications have gotten further before they proved to be dead ends. One is the use of lasers to break up rocks in the mining or digging of tunnels. Some promising results were reported in the 1960s, which indicated that shining a high-power laser on rocks could help break them up so that they could be more easily removed during mining. However, further research showed that the technique was impractical.

Finally, there are certain ideas that sound sensible only to someone who knows almost nothing about lasers. The offices of *Laser Focus* magazine, for example, once got a call from a man who proposed using a laser to trim the edge of a lawn neatly. He obviously didn't know the size of the laser that would be needed (about as big as a console television set) or how much more it would cost than ordinary lawn-edging equipment (the laser would cost $10,000). Nor could a laser stand up to the conditions tolerated by a lawnmower.

Another caller identified himself as "working in the petroleum industry" and asked where he could get a laser that could bore test holes thousands of feet into the ground. He wasn't discouraged by the initial warning that it might cost a lot. He said he was trying to replace a rig that cost $6 million to $10 million. It took a while to explain to him that it's a job lasers *can't* do for many reasons: they lack the power to drill through thousands of feet of material (a couple of inches is tough enough); there's no way to get rid of the vaporized material; and that if somehow such a laser could be built, the Department of Defense would probably classify everything and everyone in sight!

9 THE OPTICAL RULER: THE LASER AS MEASURING STICK

If you've ever tried to fix up an old house, you know how hard it was to maintain a straight line fifty or a hundred years ago. The walls look straight and seem to meet at right angles—until you try to put up wallpaper. Then you discover that the windows and door frames are subtly out of line with the walls, and that one corner of a room may be a quarter of an inch shorter than another. If the house is more than a hundred years old, the difference may be even larger. The problem is not always caused by settling, nor is it due to incompetence on the part of the builders, as is evident from the general solidity of such houses. Rather, it was the inadequacy of measuring tools that were available when the house was built.

Today, the laser helps contractors maintain straighter lines and define right angles that are really perpendicular. If a carpenter wants to draw a straight line around all four walls of a room to help him install a protective strip of molding, say, he can place a special laser instrument in the middle of the room. The laser doesn't move, but the instrument contains a prism that rotates continually, directing the beam around a full circle. The result is a thin line of laser light around the walls, far more accurate than a line drawn laboriously by hand. The red laser light also serves as a constant reference point for aligning floors and ceilings. And special optics can be added to the laser system to project a laser beam at a right angle to the reference plane formed by the rotating prism, to help the workers get the corners, windows, and doors straight.

Building straighter buildings is but one of many uses of the laser in its role as optical ruler. In the late 1960s and early 1970s, lasers began to find applications on construction sites and in machine shops. Growth was rapid, and today there are well over 100,000 lasers doing measurement and alignment jobs in construction, industry, and agriculture. Virtually all of these are low-power helium-neon lasers, emitting continuous red beams, which by themselves cost only a few hundred dollars, though complete instruments incorporating the lasers generally cost much more.

In industry, lasers help measure distances and angles, and their beams are used to draw straight lines for applications ranging from laying sewer pipes and grading soil for irrigated farming to aligning precision machine tools. Lasers can measure the shapes of parts on an assembly line and inspect them to see if they meet quality standards. They can detect subtle motion along geological faults and detect air pollutants with concentrations as small as a few parts per billion. Special types of lasers can even detect rotation and function as gyroscopes for aircraft.

Lasers are also turning into yardsticks in the hands of research scientists, who use them to make ultraprecise measurements of time, motions of the earth, and minute concentrations of chemicals. The importance of these laboratory uses of lasers was recognized in awarding the 1981 Nobel Prize in physics to two founders of laser spectroscopy. All types of lasers are used in these laboratory measurements, including instruments costing many tens of thousands of dollars.

We'll focus first on industrial uses of laser measuring sticks.

LASERS IN CONSTRUCTION

As construction projects get bigger, it's more important than ever to maintain straight lines and produce exact right angles. Sophisticated optical instruments have been developed over the years to help surveyors and construction workers make straight lines. Add a laser to such instruments, and you have a super alignment system. Virtually all lasers used in construction are small, rugged helium-neon lasers, which emit low-power red beams readily visible to either the human eye or electronic detectors.

The simplest laser used in construction is the alignment laser. In its simplest form, it just sits somewhere and projects a beam of red light that defines a straight line. To see where the line is, workers must in-

Lasers are used to control land-leveling equipment. Since such operations are often carried out 24 hours a day, the use of laser light is doubly advantageous because lasers let them be performed at night, as shown here. Courtesy Spectra-Physics

tercept the beam with a piece of paper or some other object. (Remember, a laser beam cannot ordinarily be seen along its entire length, so to detect it, some obstacle must be put in its path; workers should not look directly into the beam, because they could damage their eyes.)

More complex alignment systems have electronic receivers to detect the laser beam. These receivers produce signals that can control equipment. A bulldozer, for example, can be controlled by a laser, to make sure it cuts the right grade on a construction site. A laser set on a tripod shines its beam along the desired angle. An electronic light detector on the bulldozer's blade "catches" the beam. Automatic equipment moves the blade up or down to make sure the detector receives the beam all the time, thereby keeping the bulldozer blade plowing at the right depth. One of the commonest uses of such lasers is in irrigated farming. Farmers who want to use the least water and cause the least erosion need precise grades on their land, and automatic laser leveling works best.

Precise grades are also important in laying drainage and irrigation tubing and in digging drainage ditches. Laser alignment equipment similar to that used on bulldozers can be coupled with ditch-digging equipment to automatically control the depth of the digging.

Lasers address an age-old problem in tunneling—how to make sure the two ends meet in the middle. Laser alignment equipment helps men and machines keep to a straight line. And since you have two crews digging from either side, you have two laser lines projected. Other surveying equipment, some of which also includes lasers, can be used to define where the two straight lines are in relation to each other, and to make sure that they will ultimately meet in the middle.

We mentioned earlier that lasers indicate not only straight lines but entire plane surfaces as well, as in the case of the builder's laser, which is passed through a prism to project a beam around a full circle. Besides enabling builders to get the windows and corners in a house straight and the floors level, a laser-projected plane helps in a variety of specialized tasks common to many construction projects today, such as installing suspended ceilings, partition walls, and raised floors, such as those used in computer rooms, where cables pass under the floor. Builders who use such lasers report that productivity is increased by as much as 25 to 50 percent, that accuracy is improved, and that better use is made of manpower, because engineers and surveyors are freed for other tasks.

SURVEYING

In surveying, as in construction, defining straight lines is the name of the game. In modern instruments, the operator lines up a distant reference point through a telescope equipped with cross hairs, and the laser beam is then aimed through the same optics at the same point. To insure safety, the system is designed so that the beam can't go through the lenses while the operator is looking through them. Many such instruments can be used to measure angles as well.

Laser surveying instruments have been used for such tasks as grading swimming pools, determining differences in elevation, aligning fences and brickwork, laying out athletic fields, aligning retaining walls, grading irrigation ditches, staking out foundations, checking the pitch of pipes, setting rows of crops, leveling floors, aligning structural steel in buildings, establishing street grades, and setting sewer lines.

Another type of laser surveying instrument can quickly and automatically measure distances, even distances of many miles. Again, the operator looks through a telescope or binocularlike sight to focus cross hairs on a target at which a special mirror called a retroreflector has been set up. He then triggers a laser beam, which hits the retroreflector, which is designed to return the beam to the place where it originated—in this case to be seen by a sensor on the surveying instrument. The distance is determined by measuring the time it takes the light to make the round trip from the laser to the target and back. Either helium-neon lasers, which emit a continuous red beam, or semiconductor lasers, which fire a series of pulses or a continuous beam in the near infrared, are used in these devices. In either case, the beam spreads out enough in its travel to avoid any eye hazard.

Laser distance-measurement has proved to be a boon to surveyors, who can now measure distances as long as 65 km (40 miles) automatically and do it much faster than with conventional equipment. For example, reference points in aerial photographs have been matched with ground measurements in the Grand Canyon over an area of more than 800 sq km (300 sq miles). In order for aerial photographs to be useful in mapping, landmarks must be established. Surveyors must measure the distances between these points and match the points to the photographs, and you can imagine what a job this would be for an area as large and as rugged as the Grand Canyon, using traditional methods. As a matter of fact, it would take one year . . . and a hundred men. By contrast, two surveyors, using modern laser instruments, were able to do the job in three days! The measurements were also more accurate, because of both the inherent accuracy of the laser instruments and the elimination of the opportunity for many types of human error.

MEASURING THE EARTH'S MOVEMENTS

Lasers can also make measurements on a grander scale, to aid scientists studying earthquake prediction and other geophysical research. Earthquakes, of course, involve major movements of the earth's surface. Small, subtle motions can be warning signs, pointing to regions that are under stress or strain. Lasers can measure these subtle motions with exacting precision.

One such technique involves the Laser Geodynamic Satellite, called LAGEOS, which was launched in 1976. The satellite does not carry a laser, but its surface is covered with 426 retroreflectors. Special lasers on the ground shoot ultrashort pulses at LAGEOS, so short in fact that each pulse of light is only 60 cm (about 2 ft) long. The retroreflectors bounce the laser beams back over exactly the same path as they took to reach the satellite. When the pulses return, hardware on the ground measures how long it took them to make the round trip and where in the sky the satellite is located. Several laser systems around the globe perform the measurements, yielding data that are fed into computers. Correlation of the data tells geophysicists the relative positions of two points on the earth's surface with an accuracy of a few centimeters (an inch or so). Repeated measurements identify changes in relative position that wouldn't be detectable by other means.

Local surface motion can be detected with phenomenal accuracy simply by making measurements across faults with ground-based laser instruments. Two University of Washington researchers, George R. Huggett and Larry Slater, tested a system near two fault lines in Hollister, California, in 1975 and in only a month were able to detect movement that would have taken a year to detect by conventional methods. Their laser system was able to measure distances of 1 to 10 km (0.6 to 6 miles) with a precision of one part in 10 million—that is, with an error of only 0.1 to 1 mm (0.004 to 0.04 in).

Neither the fault measurements nor the LAGEOS program are going to enable us to predict earthquakes reliably right away, but that's not their fault. Their role is to provide highly accurate measurements of the motion of the earth, which geophysicists can then correlate with earthquake patterns. It will take years to develop a thorough understanding of how earthquakes are related to motions of the earth, but once such an understanding is developed, precise geophysical laser measurements may help pinpoint likely earthquake sites.

MICROSCOPIC LASER MEASUREMENTS (INTERFEROMETRY)

So far we've talked about measurements of distances large enough to see. But the precision of laser measurements isn't limited to such distances. By using the coherent properties of laser light—the fact that the light waves in it are marching along precisely in phase—we can measure distances shorter than a single wavelength with a technique called *interferometry*.

Waves have a special property. They can *interfere* with one another. This means that if two waves are superimposed, their heights, or amplitudes, are added to each other. Let's assume for simplicity that the two waves have the same wavelength. If they are exactly in phase, they add *constructively*—in other words, if you superimpose them, you will end up with one larger wave, whose amplitude is twice as big as the two original waves (see diagram 11). If the two waves are as far out of phase as possible, they interfere *destructively*—that is, their amplitudes cancel each other out, and the overall amplitude equals zero. If the waves are neither exactly in nor exactly out of phase, but somewhere in between, they will add up to an intermediate amplitude.

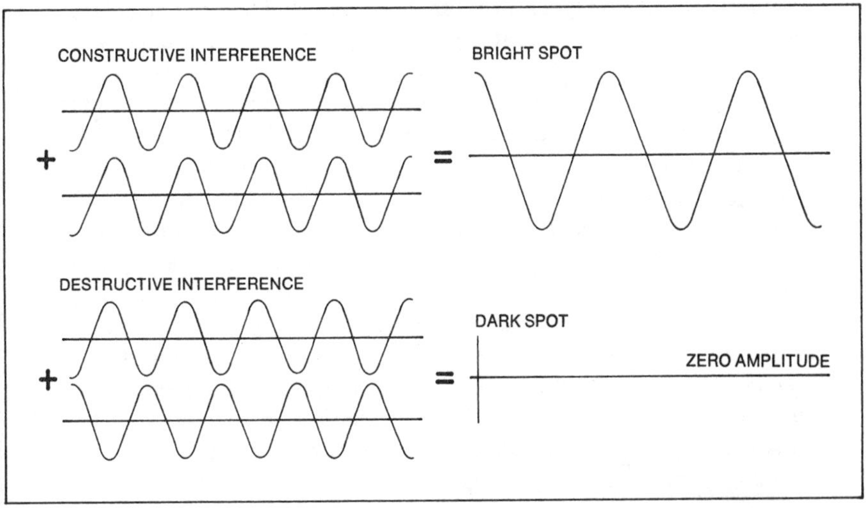

11. If you superimpose two waves that are exactly in phase, you get constructive interference, which means that the waves add together to produce a wave whose amplitude is twice as big as that of each individual wave, and a bright spot will result. If, on the other hand, you superimpose two waves that are as far out of phase as possible, you get destructive interference, which means that the two waves cancel each other out, and a dark spot appears.

With a laser, all the waves have the same wavelength and start out in phase, so any interference between two light waves from the same laser arriving at the same point over different paths depends on the difference in the lengths of the paths. If the difference is an exact number of wavelengths (say, seven), the spot is bright, because the waves interfere constructively. If the paths differ by an exact number of wavelengths plus half a wavelength (say, seven and a half wavelengths), then the spot is dark, because the waves interfere destructively.

That's the basis of an instrument called an *interferometer.* In its simplest form, it consists of a laser, a couple of mirrors, and an optical beam-splitter, which divides the laser beam into two beams of roughly equal intensity. Each of the two beams is sent off to one of the mirrors, which reflects it to a spot at which the two beams interfere. To measure distance with an interferometer, you leave one mirror fixed and move the other over the distance you want to measure. As the mirror moves, the interference spot goes from dark to light and back. Each dark-light-dark cycle indicates that the difference between the distances the two beams have traveled has changed by one wavelength of light. To measure the distance involved, you simply count the number of light-dark-light cycles, then multiply by the laser wavelength. Simple and extremely precise.

What we've just given you is a theoretical example. At present, industry more often uses interferometers to make sure that the dimensions and placement of machine tools don't vary. In such cases, interferometers compare a beam traveling along the dimension to be measured with a second beam traveling a constant distance within the instrument. At the start of the workday, the instrument is adjusted so that its sensor sees either a bright spot or a dark spot. If the sensor sees any change in the intensity of the spot, indicating that the tool has moved, it can relay the information to a control system, which automatically compensates for the motion.

Why do we need such precision? Machine tools heat up as the day goes on, and when they are heated, they expand. They also bounce around slightly. Laser interferometers can detect such subtle motions. In fact, they can detect and compensate for movements smaller than 0.001 mm (0.00004 in)—and do it automatically. That type of precision isn't always necessary by any means, but there are some situations in which it can be critical, such as the machining of components that are put together in such precision equipment as jet engines.

So far, we've only talked about looking at interference at a single spot. However, interference occurs at any point where two light waves intersect. When interference is visible over a large area, you see alternating light and dark *fringes,* corresponding to zones where interference is constructive and destructive, respectively. With the proper sort of optical arrangement, you can use interferometry to learn the precise contour of an entire surface.

To understand this capability, let's consider another hypothetical example. Suppose you want to see how flat a surface is. What you need is a laser and a piece of glass that is extremely flat. You lay the glass on top of the surface and shine the laser through it. Some light is reflected from the flat glass, but some passes through the glass and is reflected by the surface. The light waves reflected from the glass and the surface underneath the glass interfere with each other, creating a pattern of fringes, which indicates how much difference in flatness there is between the surface and the glass. If the fringes are few and are widely spaced, the surface is almost flat; if there are many, closely spaced fringes, it isn't. The contours of the fringes indicate how the surface deviates from being flat.

Although we used testing for flatness because it is a simple example, it's also one that has some practical application. Another use of interferometry involving contours is in checking whether telescope and other fine lenses and mirrors are shaped correctly. A master test piece known to be exact is placed against the lens being tested. Some light is reflected from the master; other light passes through, and is reflected by, the lens or mirror being tested. The two beams of light are reflected back and superimposed on each other to form an interference pattern. Optics were tested this way with ordinary, incoherent light for years before the laser came along, but helium-neon lasers are now generally used, because they produce interference patterns of better quality.

INSPECTION ON THE ASSEMBLY LINE

Lasers are also being used to make straightforward measurements of products coming off production lines. The simplest way to do this is to set a laser next to a conveyor belt and shine its beam slightly above the maximum allowable height of the product. Say you've got a factory that makes metal cans for a soup company, and each can must be trimmed to an optimum height of 10 cm (about 4 in).The laser beam is set to a point above the conveyor belt a hair above 10 cm. The beam hits a receiver on the other side of the belt, and any can that interrupts the beam alerts the mechanism that it is a reject.

There is a more complicated, but more precise, way of using lasers as production-line inspectors. A laser is mounted *above* the conveyor belt and shines its beam down at the tops of the parts being inspected at a fixed angle. The light is reflected and bounces up into a lens, which focuses it onto a long narrow detector that produces an electrical signal that varies according to where the light hits it. The point at which the laser beam hits the detector depends of course on the height of the component on the conveyor belt. The end result is a precise electrical signal that tells an operator or a machine the size of each product.

SCANNING CLOTH AT 55,000 MPH (88,000 KM/HOUR)

Lasers are helping to eliminate the drudgery of manual inspection of textiles. Several textile mills in North and South Carolina are using systems built by the Ford Aerospace and Communications Corporation, a subsidiary of the Ford Motor Corporation, around helium-neon lasers. The textile mills consider laser inspection faster and more accurate than human inspection.

"Toward the end of an 8-hour workshift, the little lady [inspector] just gets tired," said Robert Thompson, a spokesman for Spring Mills, Inc., which uses laser scanners at textile plants in Ft. Mill and Chester, South Carolina. The laser beam is split into three separate beams, each of which moves at 55,000 mph (about 88,000 km/hr) and scans the 40- to 64-inch-wide (1 to 1.6 m) fabric. The cloth whips by at 250 yd/min (about 4 m/sec). If any defects are found, a nozzle a few meters (or yards) from the laser squirts ink on the flaws, to aid in subsequent processing of the cloth. The laser system, meanwhile, records not only the existence of the defect but the type as well, which is important, because some defects are repairable, while others require that a piece of the cloth be cut out and replaced.

Installation of the laser system, whose overall price tag was estimated unofficially at $225,000, with an additional $75,000 required for accessories, enabled Spring Mills to cut the number of human inspectors by 60 to 70 percent. All these displaced inspectors were given other jobs in the mill—or so they say.

LOOKING FOR ROUGH EDGES

Can you imagine having a rough-edged lens implanted in your eye? Or being injected with a rough-pointed hypodermic needle? Well, it can happen, but one of the laser's new jobs is to prevent such things.

One of the heralded breakthroughs in the treatment of cataracts is the lens implant. A cataract occurs when the natural lens in the eye becomes opaque. When this happens, eye surgeons can often replace the natural lens with a plastic lens. In most operations, the surgeon fixes the replacement lens in place with sutures or clips, but in others—about a third of all such operations—the new lens rests directly against the eye tissue. If the edges of the plastic lens are the slightest bit rough, they can irritate, or even cut, the eye.

Two laser methods for detecting rough edges of lenses have been proposed. In one, two laser beams intersect at the edge of the lens, and the interference fringes in the light reflected from the edge of the lens reveal any irregularities. In the other, a single laser beam is focused on a tiny spot 0.01 mm (0.0004 in) in diameter on the edge, and the reflected light reveals the smoothness, or roughness, of the lens. Both methods are being developed by David W. Vahey of the Battelle Memorial Institute's Columbus (Ohio) Laboratories and Edward Mueller of the Food and Drug Administration's Bureau of Medical Devices.

Lasers are already at work sparing us the pain of rough edges and other defects at the tips of hypodermic needles. This inspection technique is based on the way light from a helium-neon laser beam is scattered from the tip of a needle. Vertically mounted needles pass through the laser beam, and the scattered light is observed by a special multi-element detector placed behind the needles. Output from the different elements of the special detector is processed by a minicomputer, which compares the signal from the needle being inspected with the ideal signal from a good needle. The system can inspect as many as twelve needles per second, which is much faster than humans can do it and is comparable to the rate at which the needles are produced. The laser system was developed by Recognition Systems, Inc., of Van Nuys, California.

TESTING SOLDER JOINTS

Laser inspection isn't always simply a matter of looking passively at an object. To test solder joints, a system developed by Vanzetti Infrared & Computer Systems, Inc., of Canton, Massachusetts, heats the joints with pulses from a neodymium-YAG laser. The pulses heat the joint without melting the solder. An infrared detector then watches how the joint cools. The resulting "thermal signature"—a measurement of how much infrared radiation is emitted over time—can be compared with the distinct patterns emitted by good and bad joints.

The laser technique would complement the ultrasonic inspection systems now used to test printed-circuit boards, according to the developers. In 1980 Vanzetti was building a prototype system for the Air Force Logistics Center in Sacramento.

SEWAGE AND NEWSPAPERS

Lasers can also control entire processes. For example, a laser can monitor the concentration of suspended solids in wastewater, and the information generated by the laser can be used to help control the process producing the wastewater. One common application is in sewage treatment.

A laser can serve a very different process-control function at a newspaper-printing plant, by accurately counting the number of papers printed. A simple laser system manufactured by Nosler Systems, Inc., in Eugene, Oregon, and priced at about $3,000 can count up to 30,000 to 40,000 papers without error. These systems are used by over a hundred newspapers, including the *San Francisco Chronicle,* the *Los Angeles Times,* and the *Cleveland Plain Dealer.*

LUMBER

The interest in being able to count newspapers precisely was stimulated by the soaring cost of paper and of the wood that goes into making it. Now, laser measurement techniques can help lumber mills get more useful wood out of a log than would otherwise be possible.

The simplest laser system in the lumber industry is the laser "guideline." It's simply a small helium-neon laser with a scanning mirror that spreads a streak of red light onto a log. The guideline is aligned with a saw so that the red line of laser light indicates exactly where the saw will strike the log. The saw operator uses this light to help him guide the log into the desired position. These systems are simple and inexpensive—only $800 to $1,000—and several thousand of them are in use around the world.

The lumber industry is also finding increasing use for a newer, and more complex, laser instrument—a scanner that measures the dimensions of a log and feeds the information into a computer. The computer analyzes the laser-generated measurements and calculates how to obtain the most wood from the log. The most sophisticated scanners are used on logs to be peeled for veneer for plywood. The veneer-peeling process resembles the unwrapping of a roll of paper. An eight-foot-long (2.4 m) knife peels off a one-eighth-inch (3-mm) layer of wood as the log is rotated.

Use of a laser scanner for veneer peeling can increase the yield of veneer by 5 to 10 percent over what could be obtained by the traditional process, in which the log is rotated around an axis defined by the midpoints of its two ends. Measuring the log to see what would be the best axis to rotate it about, rather than simply assuming that the log is a perfect cylinder, increases the yield. For a log 12 inches (about 30 cm) in diameter, the correction is typically only a fraction of an inch. However, the cost of wood is so high (particularly for veneer stocks), and the volume of wood handled is so large, that these new scanning systems—even with price tags as high as $200,000—are expected to save many sawmills money.

LASER GYROSCOPES

So far we've talked about how lasers can measure distance along a straight line or count or inspect things. Special types of lasers can also measure rotation. Such laser rotation sensors, or gyroscopes, are being used on the new Boeing 757 and 767 aircraft and are being considered for military aircraft and guided missiles as well.

Sensing rotation with a laser requires two beams from the same laser traveling in opposite directions around a closed path. You might think of the closed path as a circle, but in actual laser gyroscopes it's typically a triangle defined by three mirrors. The mirrors are carefully aligned to reflect the light in a complete circuit of the triangle. The active medium of the laser is along the edge of the triangle (i.e., along the path defined by the mirrors), and together the three mirrors function as a laser resonator, just like the two mirrors on opposite ends of an ordinary laser tube, as we explained in chapter 2.

It's possible to sense rotation around the closed path because the beams going in different directions see subtly different things. The mirrors move slightly as the light travels between them, so light traveling in the same direction as the rotation has to go a tiny bit further between mirrors than light going in the opposite direction. This means that one beam sees a laser resonator that's slightly longer than that seen by the other beam. Because the length of the resonator affects the wavelength of the beam, this leads to a slight difference in wavelength between the two beams. This wavelength difference manifests itself in interference effects, which can be detected and interpreted in terms of the rotation rate by a computer.

The type of rotation sensor we've described measures rotation on only one axis, but in our three-dimensional universe, there are three possible axes of rotation. Thus practical laser gyroscope systems, such as those used in the new Boeing aircraft, use three laser gyroscopes put together so that each measures a different axis of rotation. The output from such a system can be analyzed by a computer to help tell the pilot of an aircraft which way he's going and where he is on the globe.

SPECTROSCOPY: ATOMIC PROBE AND POLLUTION DETECTIVE

While the laser techniques we've talked about so far are sensitive by industrial standards, they're often crude in comparison with the ultrasensitive laser measurements made in scientific laboratories. The laser has proved to be a tremendously potent measurement tool for probing the nature of matter—a fact recognized by the awarding of the 1981 Nobel Prize in physis to two men who played key roles in the development of laser techniques, Arthur Schawlow of Stanford University and Nicolaas Bloembergen of Harvard University.

To understand the nature and importance of their work, we need to take a brief detour into atomic physics. Atoms and molecules can be identified by the wavelengths of the light they absorb or emit. The study of these wavelengths of light is called *spectroscopy,* and the goals of spectroscopy include identifying atoms and molecules and learning about their internal structure. Lasers can play a vital role in these studies. By tuning the output wavelength of lasers as they illuminate samples of material, scientists can learn about the constituents of those samples. They can monitor the amount of laser light absorbed, the production of light at other wavelengths, or other light-induced phenomena, such as ionization of atoms or molecules.

Laser spectroscopy quite literally casts new light on what goes on inside atoms and molecules. The way in which light is absorbed and emitted by atoms and molecules indicates how energy is distributed within them. Changes in absorption or emission tell scientists how atoms and molecules react to changes in external conditions. Although the effects are often subtle, the laser is up to the challenge of detecting them. For example, lasers can readily detect the slight splitting of spectral lines caused by a static magnetic field. Laser spectroscopy can also observe many other interactions involving atoms or molecules, revealing new information about the nature of matter. The details are far too complicated to go into here. Suffice it to say that laser spectroscopy has taught scientists much about such things as the interatomic processes involved in chemical reactions. Such understanding can ultimately have important payoffs in improved techniques for chemical processsing.

Although Schawlow and Bloembergen, both in their early 60s, have known each other for many years, they have never worked together on a research project. They made their contributions to fundamentally different but complementary subfields within laser spectroscopy. Schawlow was a founding father of *linear* spectroscopy, while Bloembergen has been the central figure in the study of *nonlinear* interactions of light and matter. The best way to explain the difference is to look at the more familiar example of an audio amplifier. When the amplifier's output is undistorted, its linear—directly proportional to the input (perhaps a factor of 20 higher, for example). When the output is distorted, its nonlinear—related to the input in a more complicated way. Nonlinear effects may sound bad in an audio amplifier, but in spectroscopy they can complement linear techniques by revealing interactions that otherwise would not be detectable.

Spectroscopy makes a fascinating laboratory plaything for physicists and chemists, but it can also serve practical purposes. Prominent among these is the monitoring of air pollution. For example, certain lasers can be tuned to a wavelength at which sulfur dioxide, a pollutant produced by burning high-sulfur coal and some other fuels, has a strong absorption line—that is, a wavelength at which it absorbs light strongly. If a laser beam is passed through a sample of air, the amount of light absorbed at that wavelength by the air gives a measure of how much sulfur dioxide it contains. This method normally requires a detector on the other side of the smokestack being tested, but new, re-

This pollution-measuring system includes a laser that sends light into the atmosphere and a telescope that collects light scattered from the pollutants being measured. Courtesy SRI International

mote-monitoring techniques have been developed that require no detectors at the actual site. They rely on the *scattering* of light from particles in the air to return signals to an observer at the same site as the laser. The strength of those signals is a measure of the concentration of the pollutant. The laser detector can be aimed at suspected polluters without their knowledge. The Environmental Protection Agency has prototype laser pollution monitors, but at this writing, they are not yet in widespread use. Remote laser monitors are also being tested in Europe.

A principal advantage of laser techniques is their sensitivity—reaching beyond the parts-per-million into the parts-per-*billion* range. In the laboratory, laser techniques have gone even further—to detection of a single atom of cesium. While laser techniques are not being used routinely for measuring low concentrations of atmospheric pollutants, they are proving themselves in the lab. The list of research measurements is long and impressive. Laser experimenters have measured pollution-level concentrations of sulfur dioxide, sulfuric acid, nitrogen oxides, and some hydrocarbons. Laser measurement is also being used to study automotive exhaust at General Motors Research Laboratories (but not yet at your local gas station). GM has measured carbon monoxide, methane (the simplest hydrocarbon), and sulfur dioxide in car exhaust, as part of its studies of combustion and catalytic conversion.

Spectroscopy may turn out to be a double-edged sword, says Arthur Schawlow, and end up being as much of a social and philosophical issue as a scientific one. Schawlow points out that once lasers can identify single atoms or molecules of *any* substance, then carcinogens will be readily detected in our foods. What are the chances, he recently asked us, that you wouldn't find a single molecule of a cancer-causing agent in, say, a quart of milk? Obviously, it's likely that any reasonable quantity of food will have a single, stray molecule of a harmful agent in it. What do you do with such knowledge? The laser may eventually give us a precise accounting of what we're eating, drinking, and breathing. And the real question may turn out to be: Do we really want to know?

LASER TIMING

Finally, lasers are becoming the ultimate synchronizing devices for clocks around the world.

It was the development of the transcontinental railroad that first made it necessary to synchronize time zones across the United States. Since then, the development of sophisticated telecommunications systems has made it necessary to synchronize clocks around the world.

Today, radio links maintain day-to-day international synchronization of clocks to within 100 billionths of a second (100 nanoseconds). A laser-satellite approach proposed by the Bureau International de l'Heure, a Paris-based agency that maintains international time standards, could improve that accuracy to within one billionth of a second (1 nanosecond).

The idea is to aim several ground-based lasers at a satellite receiver. The lasers emit short pulses at precise intervals, and ground-based equipment times how long it takes for the pulses to be returned by retroreflectors on the satellite. At the same time, a receiver on the satellite detects the signals and measures precisely when they arrive at the satellite. The various lasers are supposed to aim pulses at the satellite at precise times; the satellite receiver notes the difference between when the pulses were supposed to arrive and when they did arrive, and that information is used to synchronize the clocks driving the lasers.

Experiments to test the concept were to begin in mid-1981 with the launch of the Sirio-2 satellite by the European Space Agency. In the experiments, called LASSO for Laser Synchronization from Stationary Orbit, lasers at fifteen ground stations in eleven countries fire pulses triggered by cesium atomic clocks. The satellite is geosynchronous—that is, its orbit precisely equals the earth's rotation period, which means that it appears to stay stationary above one point on the earth. It was to be stationed first over the Atlantic Ocean, so that stations in North and South America could participate in the tests, and then moved to a position over Europe and Africa for an ocean communications experiment that is one of the satellite's other functions.

The experiments are scheduled to last for only two years, and the system isn't designed for all-weather operation, since the laser pulses can't penetrate clouds. But even before the satellite was launched, researchers at the European Space Agency and the French National Space Center in Toulouse were designing future experiments that would improve synchronization accuracy to one-tenth of a billionth of a second (0.1 nanosecond). Such synchronization could pay dividends in improving the precision of navigation, positioning, and timing of celestial events, as well as aiding in international telecommunications.

10 ZAPPING ENERGY FOR THE TWENTY-FIRST CENTURY

So you want to produce energy with a laser. Start with a gold-plated glass balloon no bigger than a grain of sand and mount it on the tip of a needle-like positioning device. Inside the balloon, or pellet, is the fuel for fusion: a mixture of deuterium and tritium, two rare isotopes of hydrogen. The gold plating around this tiny balloon allows it to absorb energy more efficiently.

Then aim a laser at the balloon, which is really a target. Or to be more precise, move the target so that the laser beam can hit it. It isn't very easy to move the laser, you see. It fills up most of a four-story building at the Lawrence Livermore National Laboratory in Livermore, California, and it's mounted on a massive framework of girders, so that even an earthquake won't move it very much. The laser cost about $25 million to build, and it can produce a pulse lasting only a billionth of a second that is more powerful than any other laser in the world. It's called Shiva, after the multiarmed Hindu god of both creation and destruction, and it's a single laser system whose beam is split into twenty separate beams, each of which is amplified several hundredfold before being focused into a chamber in a spherical array of light that converges upon the tiny target.

A gang of microprocessors adjusts the twenty beam paths to make sure they will all converge on their sand-size target, a task that has been compared to hitting a basketball with a shotgun at a distance of 18 miles (29 km). When Shiva lets loose its fury, the glass-and-gold shell of the pellet evaporates and the core is compressed so much that the deuterium and tritium nuclei fuse together under pressure a hundred billion times that of the earth's atmosphere. The result: an explo-

In this photograph, we see six of the twenty Shiva fusion laser amplifier chains. The Shiva laser system can deliver nearly 30 trillion watts of power to a target the size of a grain of sand in 0.2 billionth of a second. Courtesy Lawrence Livermore National Laboratory sion vastly smaller than, but otherwise not unlike, that of a hydrogen bomb, which releases energy in much the same way as energy is produced at the core of the sun and other stars.

This is laser *fusion,* and it holds out the hope of extracting nuclear power from a common substance we don't ordinarily consider a fuel at all—seawater. What we've just described *has* happened, but it's just a beginning; practical laser fusion energy isn't right around the corner. The Shiva laser is in a laboratory, not a power plant. At present, much more energy has to be put into making the reaction go than we can get out of it. Obviously, the goal is to reverse that situation. And frankly, progress has been slow and sometimes disappointing. But some scientists at Livermore and several other labs believe that lasers could provide the key to harnessing thermonuclear fusion by the first years of the twenty-first century.

In this chapter we will focus on the promise of laser fusion. We will also look at two other areas in which lasers may contribute to our energy needs: the enrichment of uranium, in which lasers can help provide the right mixture of uranium isotopes for use in conventional nuclear reactors based on *fission,* and a far-out plan to use laser beams to bring solar energy back to earth from satellites equipped with solar collectors.

You'll find a common subtheme running through our examination of laser energy: the military implications of each technology. Laser fusion simulates the explosion of a hydrogen bomb. Lasers could enrich uranium for atomic bombs as well as civilian nuclear power plants. And a laser beam powerful enough to transmit energy from a solar power satellite system would also have plenty of energy for less-peaceful purposes (though it wouldn't have all the critical beam-direction hardware needed to make it an effective weapon).

We'll begin with fusion.

THE LURE OF FUSION

Tremendous energy lurks within the atomic nucleus. We know two ways to get it out: fission and fusion. The first to be discovered and the first to be tamed was fission, which is the splitting of the nucleus of a heavy atom, like uranium, into two lighter nuclei. In this reaction, energy is released. Fusion is the opposite process. It is the merging of the nuclei of two smaller atoms, such as hydrogen atoms, to yield a larger nucleus . . . and energy.

In practice, we generate fission energy by splitting atoms of uranium or plutonium in a nuclear reactor or an atomic bomb. Strictly speaking, both "nuclear" and "atomic" could refer to fusion as well as fission. We'll use them only to refer to fission, however, following the general practice. Fusion we will always call "fusion," and nothing else. It's just another example of technology moving faster than language.

Fission has problems. It produces radioactive waste, and fissionable fuels—uranium, plutonium, and sometimes thorium—will run out within a few centuries. And that's an optimistic estimate.

The basic raw material of fusion is hydrogen, ample supplies of which are available in ocean water. The sun burns ordinary hydrogen, which has a nucleus consisting of a single proton. This is the isotope hydrogen-1. A man-made fusion reactor would burn deuterium, or hydrogen-2, which is a heavier isotope. It's the same basic element, but its nucleus has both a proton and a neutron. Deuterium accounts for only 0.015 percent of natural hydrogen, but there's still plenty in the oceans (each water molecule contains two hydrogen atoms and an oxygen atom). The first fusion reactors are likely to burn a combination of deuterium and tritium, such as was contained in the gold pellets we described earlier. Tritium is hydrogen-3, with a nucleus of one proton and two neutrons. Tritium does not occur in nature, because its nucleus is unstable and decays in about a dozen years. However, it can be produced readily by using neutrons to irradiate lithium, a light soft metal that is much more abundant than uranium (there are more than 500 lithium atoms in the earth's crust for each uranium atom).

In theory, fusion could be made a relatively clean process. That goal is a long way off, however, as the fusion reactions most likely to be practical in a reactor all produce radioactive tritium or free neutrons.

Fusion requires very high temperatures and pressures. In nature, fusion occurs in the cores of stars, such as our sun, and provides their energy. Man has made fusion happen in the hydrogen bomb, sometimes called the thermonuclear bomb. What many people don't realize is that an H-bomb contains an atomic (fission) bomb inside it. The A-bomb's explosion provides the energy necessary to trigger the higher-energy explosion of the hydrogen fusion bomb. This is a terrific arrangement for destroying things but is not so good for providing energy for peaceful purposes.

TAMING FUSION

Ever since the explosion of the first hydrogen bomb, physicists have sought to control fusion reactions so that the energy they generate can be used for constructive purposes.

The central problems in any type of fusion are to heat the plasma or gas and compress it to the point where a fusion reaction can occur. The fuel must also be held long enough. There are two main approaches.

The oldest is called *magnetic confinement.* The term is somewhat self-explanatory; it seeks to confine the fusion reaction by applying a magnetic field. The fusion reaction itself takes place in a hot gas—actually a plasma, in which all the electrons have been stripped from the atomic nuclei. In this method physicists are basically trying to trap the plasma in a magnetic "bottle" generated by huge magnets. Unfortunately, the hot plasma is extremely difficult to contain and leaks from the bottle. In theory, magnetic confinement is not quite as hard as trying to wrap Jello with rubber bands, but in practice, it's a pretty good comparison.

Back in the 1950s, magnetic confinement seemed to be the logical approach. But the slow progress of the research, together with the development of the laser, stimulated interest in another approach, called *inertial confinement.* The idea of inertial confinement is similar to that of a hydrogen bomb: hit the fuel with large amounts of energy to heat it to high temperatures and compress it to awesome densities. In an H-bomb, the job is done by an A-bomb. In inertial-confinement fusion, a beam of energy (usually from a laser), zaps a pellet containing deuterium and tritium gas. The process gets its name from the fact that it is an *inertial* force that compresses the fusion fuel, because the inertia of the material going inward from the surface of the target is what compresses the nuclear-fuel core.

In laser fusion, a very short pulse from a high-power laser is spread out over the surface of the target pellet, which, by the way, doesn't *have* to be gold—other coatings are also being tested. If the laser energy is spread uniformly enough over the entire surface of the pellet, the heating it produces causes the pellet to *im*plode—to burst inward. The exact mechanism of the implosion depends on the design of the pellet. In the simplest case, what happens is that material is evaporated from the surface of the pellet, and the outgoing material generates an inertial force that compresses the remaining material, causing an implo-

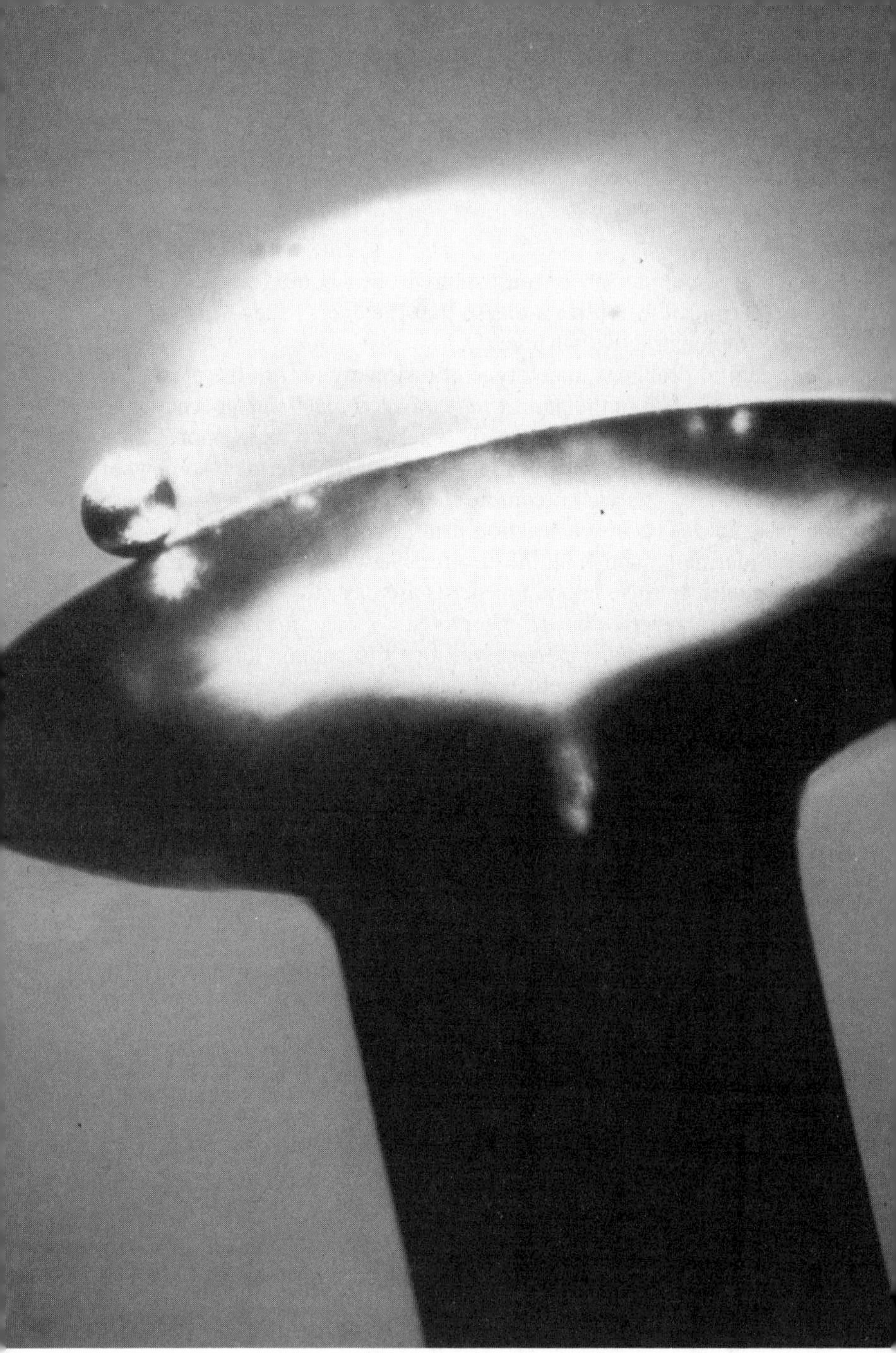

A metal imitation of a fusion pellet sits on the head of a pin. The diameter of the pellet is about that of a human hair. Courtesy Lawrence Livermore National Laboratory sion. In some more complex targets, only recently declassified because of their resemblance to hydrogen bombs, much of the implosion energy comes from X rays produced by the part of the pellet heated to extremely high temperatures by the laser.

Obviously, there are significant differences between magnetic- and inertial-confinement fusion. The crucial ones stem from how the two methods satisfy what's called the *Lawson criterion.* According to this criterion, in order for fusion to take place, the density of the plasma multiplied by the confinement time must exceed a certain value, which depends on the type of fuel used.

For magnetic-confinement reactors, the confinement time is expected to be relatively long—about 0.1 sec—so that the plasma density would have to reach 10^{15} particles per cubic centimeter (about 10^{16} per cubic inch) for the simplest type of fusion reaction to occur. That's roughly equal to the density of air 120 km (75 miles) from the ground. For laser fusion, confinement times would be extremely short—something like 10^{-9} or 10^{-10} sec—making it necessary to reach plasma densities of 10^{23} to 10^{24} particles per cubic centimeter (10^{24} to 10^{25} per cubic inch), comparable to the density of water, and around 10,000 times greater than the normal density of a gas. In other words, magnetic confinement aims at long containment of low-density material, whereas laser fusion's goal is brief confinement of dense fuel.

There are trade-offs with either process, and with both, fusion is difficult to attain. At this writing, magnetic-confinement fusion continues in the lead, at least in part because the Department of Energy (DOE) regards the older and larger program as its main hope for fusion power.

MILITARY INTEREST IN FUSION

Fusion research commanded about two-thirds of the roughly $300 million that the U.S. government spent on laser-energy research in fiscal 1981. Frankly, laser fusion's budget is justified by its military applications not by its potential to produce energy. Although the laser-fusion budget is administered by the Department of Energy, it is funded by DOE's division of military applications, and that portion of the budget originates within the military, not within DOE itself. DOE spokesmen emphasize the potential applications of laser fusion in civilian power generation, but the fact is that so far almost all federal funding for laser fusion has been justified purely on military grounds.

Why would the military care about fusion? Because the implosion of a laser-fusion target is similar to the explosion of a hydrogen bomb. The physics of laser fusion is similar, albeit on a smaller scale, to the physics of H-bombs. Target design has much to do with bomb design, and in fact some of the leading target designers have previously designed hydrogen bombs.

Atmospheric testing of full-scale hydrogen bombs is prohibited by the Nuclear Test Ban Treaty, but there is nothing to prevent scientists from testing bomb designs in a laser-fusion laboratory. That's one of several reasons why two of the largest centers for laser-fusion research are the government's two most important nuclear-weapons laboratories—the Lawrence Livermore National Laboratory in Livermore, California, and the Los Alamos National Laboratory in Los Alamos, New Mexico.

Besides simulating the explosion of a hydrogen bomb, laser fusion can simulate the *effects* of a bomb. Fusion generates an intense burst of neutrons and other subatomic particles, and military researchers need to know how various materials react to such a flux. They can use the nuclear radiation from a laser fusion reaction to test the "radiation hardness" of military equipment, without having to explode a full-scale bomb. Scientists subject the materials used in tanks and other equipment to small-scale fusion to find out how they'll stand up in a nuclear attack.

Because of these military applications, some laser-fusion technology remains classified. Target design, because of its similarity to bomb design, has traditionally been the most sensitive area. Indeed, because of the sensitive nature of some target designs, they're tested only at Livermore and Los Alamos, where stringent security requirements can be met. Declassification comes slowly. The X-ray-induced implosions we mentioned earlier, for example, didn't emerge from under security wraps until a November 1980 conference conducted by the American Physical Society, although researchers at Livermore had been studying them for many years. And even after the meeting, diagrams of such targets remained classified.

Although the funding for laser fusion has come from the military, it's clear that the prospects for developing a civilian power source have received more than lip service. DOE has sponsored conceptual designs of laser-fusion power plants and research into some of the technology that would be required to put laser-fusion reactors on line for generating power in the twenty-first century.

ORIGINS OF LASER FUSION

The concept of inertial-confinement fusion had its roots in the 1940s research that led to the hydrogen bomb. The idea of driving fusion reactions by *non*nuclear explosions apparently originated at the same time, or slightly later. The idea of laser fusion appears to have sprung up about the same time as the laser, but its true origin remains shrouded in a cloud of secrecy and conflicting claims.

We know some of the early players in the game. There was an active group at the Lawrence Livermore National Laboratory in the early 1960s, whose members included, among others, Edward Teller, generally considered the father of the American hydrogen bomb, and two researchers who remain active in Livermore's laser-fusion program today: Raymond E. Kidder and John A. Nuckolls. Then there was Friedwart Winterberg, a German-born scientist who settled at the University of Nevada's Desert Research Institute in Reno, where he worked on theoretical studies independently, without heed to security restrictions—somewhat to the displeasure of the government. Some Russians also showed interest. And of course Gordon Gould at least considered the idea of using lasers to induce thermonuclear fusion, which may—or may not—be covered by his broad patent on laser applications, a thorny question covered in chapter 4.

The government program continued at a lazy pace throughout the 1960s. During this time the program was conducted under the auspices of the Atomic Energy Commission (AEC). It eventually passed to the Energy Research and Development Administration (ERDA) in the 1970s and finally to DOE. But it was in 1969, under the AEC, that laser fusion got its first good kick in the rear.

THE KMS ADVENTURE: FAST NEUTRONS AND SUDDEN DEATH

In that year, laser fusion looked as if it would follow the path of many government research programs, with long and careful work quietly pursued by scientists in various laboratories accompanied by periodic battles among program directors, government officials, and politicians in the background. Most of the work was covered with a thick layer of security classifications. But that was also the year that something happened that would eventually drag the government program—sometimes kicking and screaming—into the limelight. Keeve M. Siegel discovered laser fusion.

Kip Siegel was a pushy, pudgy, energetic man, and he'd been very busy during the 1960s. He'd started the decade as a professor of mathematics at the University of Michigan in Ann Arbor. He'd left the university in 1960 to found the Conductron Corporation, which specialized in radar, optics, and electronics. Conductron became a successful business and was merged eventually with the McDonnell Aircraft Corporation (now the McDonnell Douglas Corporation). When he left Conductron in 1967, after a dispute with McDonnell chairman James S. McDonnell, Siegel had parlayed his original $12,000 investment into $4 million.

Within a few days of leaving, he was back in business again—this time as head of a new company, KMS Industries, Inc. (Not being a modest man, he used his own initials for the company name.) At KMS his approach was different. Instead of building a single company, he started acquiring smaller companies (paying for them with KMS stock) to build a conglomerate empire. The acquisitions were generally old companies, which he infused with new technology. The growth was rapid. In its first year, KMS's sales soared from zero to $13 million. In the second year, they jumped to $59 million. Through 1969, KMS acquired thirty-two companies, ranging from book publishers to optics makers.

It was at this point that Siegel became captivated by the allure of laser fusion. He didn't have delusions that he could match the government's immense resources. But he did think that a small group of bright, highly motivated scientists could get better results than researchers working in government laboratories. He thought that private industry could develop the key concepts for laser fusion and make money doing it. It's an old-fashioned idea, and today it has the ring of science fiction to it.

The technology was still under security wraps, and KMS had to get permission from the AEC to conduct laser-fusion research. It took a lengthy series of negotiations to secure that permission, but in 1971,

KMS was finally given the go-ahead to use its own money to study laser fusion.

KMS was able to get some support for its laser-fusion work from other companies, notably the Texas Gas Transmission Corporation and the Burmah Oil Corporation. But most of the money for KMS's laser-fusion research was raised by selling off the company's other businesses. Millions of dollars were poured into laser-fusion research by KMS and its industrial partners.

Siegel realized only part of his vision. The company found itself in a race with the government's two major laser-fusion programs, at the Lawrence Livermore National Laboratory and the Los Alamos National Laboratory. The goal was to be the first to demonstrate fusion convincingly. The evidence sought by each program was the production of energetic neutrons, and the race became known as the "neutron derby."

KMS won in early 1974. A few neutrons had been observed in earlier experiments elsewhere, but it was never clear if those neutrons were due to fusion or other processes. Using laser beams to implode deuterium-tritium pellets, the first KMS experiments produced 300,000 neutrons of the type that would be expected from fusion. Within a few months, using laser pulses delivering only 60 percent more energy on target, KMS was able to produce seven million neutrons.

However, that race was only the first leg in a marathon that still has many miles to go before laser fusion can be considered practical. By early 1975, the money started running out. KMS asked the Energy Research and Development Administration for financial support, and in March 1975 Kip Siegel went to Washington to tell Congress why he should be funded.

In the midst of his testimony, the usually persuasive Siegel had trouble getting his words out. He stopped to sip a glass of water, then was heard to mutter the word "stroke" before collapsing. The next day he was dead, at the age of 52.

KMS managed to secure a series of short-term government contracts for fusion research, but that didn't end its problems. While the KMS Fusion, Inc., subsidiary, which performs the laser-fusion research, was able to continue operations, the parent KMS Industries ran out of money in early 1976 and effectively closed its doors while looking for new investors.

It was 1978 before the company finally got on an even footing. It managed to negotiate a long-term contract with the Department of Energy, alleviating the chronic uncertainty that plagues holders of short-term government contracts. It also found a group of investors who put over $1 million into KMS Industries in return for an initial 20 percent of the company's outstanding stock. However, it took a while to iron out one problem that disturbed DOE—the fact that the head of the investment group was John E. Long, an Edmonton, Alberta, investor. DOE wasn't happy about having sensitive laser-fusion research performed by a company headed by a Canadian citizen. To soothe government fears, Long's brother Patrick, a Chicago lawyer with United States citizenship, became KMS chairman.

SHIVA, NOVA, & HELIOS

KMS still holds patents on laser-fusion concepts and still performs important research on laser fusion for the Department of Energy. But today the two largest laser-fusion programs are at the Lawrence Livermore National Laboratory and the Los Alamos National Laboratory. Each has a large laser in operation for laser-fusion research, and each is in the process of building a still larger laser in an attempt to answer the critical question of how laser fusion "scales" with increasing power—that is, how much is gained by using larger lasers to produce laser fusion.

Livermore's program is headed by John Emmett, an influential, but sometimes controversial, physicist who's been a driving force in keeping laser-fusion research going. Livermore is using neodymium-glass lasers to implode fusion targets. The neutron yields long ago passed those of the early KMS experiments (which also used neodymium-glass lasers), and scientists no longer measure results of fusion experiments simply in neutron yield.

The biggest laser in operation at Livermore is Shiva. Each of its twenty arms is actually a line of laser amplifiers that builds up the energy in a laser pulse. A single pulse from a precise laser oscillator is split into twenty pieces, and one of those pieces is fed into each of the twenty parallel amplifiers. The arms are precisely aligned and adjusted (with a computer-based system) so that the pulses passing down each arm arrive at the target at precisely the same time. The twenty arms each illuminate a separate segment of the target, in an effort to provide the uniform target illumination required for a symmetrical implosion.

Shiva delivers a pulse of about 10,000 joules to its target. That's equivalent to only about 0.003 kilowatt-hour of energy—or the energy required to operate a 60-watt light bulb for 3 minutes. However, it's delivered in about one-billionth of a second to a target only a few tenths of a millimeter (about a hundredth of an inch) in diameter. The highest power that Shiva can produce is nearly 30 trillion watts, although that's in a pulse only 0.2 billionth of a second long. By comparison, that peak power is 30,000 times the electrical output of a large nuclear power plant—but remember that the power plant can put out that much power continuously, while the laser can do so for only a fraction of a billionth of a second.

Livermore has already begun work on its next laser, Nova. It's an ambitous project, which, when completed, will have cost some $200 million. Two stages are planned. In the first, ten arms of Nova will be built, along with offices and laboratory space; the laser is to produce pulses of about 100,000 joules lasting about three-billionths of a second. Once this system is working, another ten arms are to be installed in the building now housing Shiva. Nova will produce a total of 200,000 to 300,000 joules in pulses of about three-billionths of a second when this second stage is completed in the mid-1980s.

At Los Alamos, the emphasis is on carbon dioxide lasers, rather than neodymium-glass lasers. Los Alamos is already operating Helios, a large carbon dioxide laser that produces pulses with about the same amount of energy as Shiva and has eight parallel arms. The peak power is somewhat less than Shiva's, because the pulses are longer.

In 1983, Los Alamos plans to begin operating a larger carbon dioxide laser named Antares, after the red supergiant star that is one of the twenty brightest stars in the summer sky. The massive laser will produce pulses of 40,000 joules lasting one-billionth of a second, corresponding to a peak power of 40 trillion watts. That energy will be delivered by twenty-four separate beams.

OTHER FUSION PROGRAMS, AT HOME AND ABROAD

The Livermore and Los Alamos programs are the country's largest, but there are others in addition to these and the program at KMS Fusion,

A workman balances on the end of a tube that will deliver 12 of the output beams of the Antares laser system now under construction at Los Alamos. The completed Antares will include two such tubes delivering a total of 24 beams. Courtesy Los Alamos National Laboratory

Inc. Inertial-confinement fusion using beams of ions is being studied at Los Alamos; Sandia National Laboratories in Albuquerque, New Mexico; the Argonne National Laboratory in Illinois; and the Lawrence Berkeley Laboratory in Berkeley, California. Laser fusion is being studied at the Naval Research Laboratory in Washington, D.C., and at the University of Rochester's Laboratory for Laser Energetics in Rochester, New York. Each of the laser-fusion laboratories has a large laser with which fusion experiments are performed. All receive some support from the Department of Energy. Security restrictions are imposed on all but the program at the University of Rochester, which derives much of its support from industry and is explicitly intended to permit experiments with high-power lasers by nongovernment researchers, without security impediments.

Laser-fusion research is far from unique to the United States. There are significant programs under way in at least six other countries: Britain, France, Japan, Australia, China, and the Soviet Union. Much, but not all, of the research in Britain and France is military work, which remains under security wraps, whereas in Japan and Australia, most research appears to be sponsored by civilian agencies.

The Chinese effort dates back to 1965 and is particularly impressive considering how little contact China had with the outside world for many years. The Chinese program at the Shanghai Institute of Optics and Fine Mechanics is only a few years behind the most advanced programs in the United States, according to Stephen E. Bodner and Barrett H. Ripin, two leaders of the laser-fusion program at the Naval Research Laboratory who visited China in 1980. The Chinese don't have the supporting industries that other researchers can draw on, and much of their equipment is home-made. But they're working hard and learning fast, making changes as they go.

The Soviet inertial-confinement fusion program is something of a puzzle. There are two large glass lasers at the Lebedev Physics Institute in Moscow, one for each of the Institute's two Nobel laureates, who were honored for helping develop the laser. Nobelist Nikolai Basov's group has fired its laser, and some experimental results have been reported in the Russian literature. But laureate Aleksander Prokhorov's group had yet to get its laser up and running as of early 1980, according to American scientists who were visiting at the time.

The biggest part of the Soviet inertial-fusion effort appears to be de-

A workman performs maintenance on the target chamber of Helios, the world's largest carbon dioxide fusion laser. The fusion target, on which Helios's eight beams focus, is held in the center of the chamber. Courtesy Los Alamos National Laboratory

voted to an unusual approach using beams of electrons or ions. Instead of directing the beam of electrons at a target, the beam is passed through a cylindrical metal foil, and the electromagnetic forces caused by the beam cause the foil to implode. American scientists doubt that the imploding-foil approach could be used to generate electricity commercially, but researchers at Sandia are studying imploding foils for military experiments. Although the Russians give the usual lip service to power generation, it's probable that much of their research is directed toward military applications.

THE "PERFECT" FUSION LASER

A hot controversy currently rages over what kind of laser is best for fusion. The major technical problem is finding a laser that can produce not only high enough powers, but high powers at a wavelength that can drive an inertial-confinement implosion. The solution is not obvious.

The carbon dioxide lasers at Los Alamos emit infrared light at 10.6 micrometers, and it's not clear whether light of that wavelength can transfer its energy efficiently enough to the plasma produced when the beam hits the fusion target. That's a question that Antares is designed to answer. Carbon dioxide lasers are efficient and well-developed, but if they can't transfer their energy to the target well, they won't be of much use.

The problem is that carbon dioxide lasers produce relatively long wavelengths, and shorter wavelengths are believed to work better on fusion targets. The neodymium-glass lasers at Livermore produce light at 1.06 micrometers, in the near-infrared range, close to the visible region. For experimental purposes, it's even possible to divide the wavelength of that 1.06-micrometer light by a factor of two, three, or four, by passing the light through special crystals.

The implosion of a deuterium-tritium pellet, which we described at the start of this chapter, was done with a neodymium-glass laser. But remember that this was only an experiment. Neodymium-glass lasers do not appear to be suitable for laser-fusion power plants, where they would have to fire their beams again and again and again. These lasers are very inefficient and take hours to cool off between pulses—to say nothing of being grossly expensive in the large sizes required.

So a new type of laser is needed, which some researchers have dubbed Brand X. Despite considerable research, no one has come up with one yet. Indeed, the Department of Energy is stressing that its research is directed at inertial-confinement fusion, not simply laser fusion. This means lasers are facing stiff competition from ion and electron beams, which can also be used to drive inertial-confinement implosions. (Livermore's John Nuckolls has even proposed hitting fusion targets with pellets.) Support for the ion-beam program is growing, and Lawrence Killion of the DOE told a 1980 conference on inertial-confinement fusion: "We can no longer regard lasers as the dominant option." A threatening statement, coming as it did from the director of the laser-fusion division in DOE's office of inertial-confinement fusion.

TARGET: IGNITION

The reason for such a threatening statement is a continuing series of technical problems with laser fusion, which has led the DOE to back off from what was originally to have been the next milestone for the laser-fusion program: demonstrating scientific break-even—that is, producing an amount of fusion energy equal to the amount of laser energy that hits the target. Instead, the DOE has redefined that next milestone as "ignition," a less demanding requirement to obtain the *density* needed for scientific break-even, without having to heat the plasma to as high a temperature as would be needed for scientific break-even.

Break-even is a critical goal. Right now, lasers must fire more energy into the target than comes out of it. It doesn't take a genius to realize that the result consumes, rather than produces, power. The problem is that the amount of laser energy expected to be needed for scientific break-even "has increased faster than the value of gold," in the words of John A. Nuckolls, head of theoretical physics in Livermore's laser-fusion division and a top target designer. Nuckolls made his statement at a February 1980 meeting, when the price of gold had reached a historic high, and we should point out that in the weeks that followed, the price of gold plummeted. Unfortunately it was not an omen. Predictions of the laser energy required for scientific break-even stayed within a broad range centered on 300,000 joules. Earlier predictions of break-even energy had been as low as 1,000 joules.

Nuckolls noted another problem at the 1980 meeting: the inability of the various laboratories studying inertial-confinement fusion to work together. "Inertial-confinement fusion doesn't have a chance of succeeding unless the leaders of the program learn to overcome this problem," he said. One particular problem is the tendency of some laboratories to try to gain support from Congress for their programs by criticizing the work being done at other laboratories. That problem reflects a continuing budget crunch squeezed still further by the Reagan Administration's proposed cuts, which were still pending when this book went to press.

The inflation of the energy required to achieve break-even may mean that Livermore's planned Nova system will fall just short of attaining scientific break-even (nobody's quite sure yet). However, Nova should reach the less demanding ignition level.

The next problem is that after ignition is attained, many of the program's military objectives are expected to have been completed. At that point, the military might decide to build a special laser-fusion lab for its own purposes. Efforts to develop laser fusion as a civilian energy source will then have to sink or swim by themselves. At that stage, if the program continues beyond ignition and scientific break-even, the next milestone is engineering break-even—the point at which the total energy from fusion in the pellet equals the total energy used to drive the laser, which is much more energy than actually comes out of the laser, because lasers are so inefficient. Laser developers would be delighted with efficiencies of a few percent in fusion lasers, but particle-beam developers expect to attain overall efficiencies of 15 to 25 percent in particle-beam drivers for inertial-confinement fusion.

THE FINAL STEP: A POWER PLANT

The next step after engineering break-even would be some sort of prototype fusion reactor, which would generate small quantities of energy, probably in the form of electricity. Only after all these stages have been gone through would the scene be set for construction of the first full-scale laser-fusion reactor, probably no sooner than the twenty-first century.

Designers have already given some thought to the form that a laser-fusion power plant would take. It would probably contain liquid lithium to transfer energy and breed tritium, the radioactive hydrogen isotope that is expected to be one of the fuels in the first generation of laser-fusion power plants. It would have to withstand repeated explosive shocks and be able to rapidly remove from the target chamber the debris from one explosion, so that the next pellet could be imploded.

It will be a massive engineering challenge.

URANIUM ENRICHMENT

Fusion is clearly the most dramatic use for lasers in energy. But there's a much more immediate prospect lurking in the shadows of the larger fusion programs—the production of the special, *enriched* uranium required for nuclear-*fission* reactors.

Although in theory energy could be produced by splitting any type of uranium atom, in practice only one isotope is split—uranium-235.

That isotope is rare, making up only 0.7 percent of natural uranium; virtually all of the rest is uranium-238, which does not split when struck by a neutron under the conditions found in a nuclear (fission) reactor.

The problem is that the concentration of uranium-235 in natural uranium isn't high enough for the "light-water" reactors used in the U.S. and most of the rest of the world. Such reactors require uranium-235 concentrations of 3 to 4 percent in order to sustain a *chain reaction* involving the continual splitting of uranium-235 nuclei to produce neutrons, which in turn split more uranium-235 nuclei and produce more neutrons, *ad infinitum.* Even higher uranium-235 concentrations are required for bombs. (Natural uranium can be used in the CANDU reactor developed in Canada, which uses heavy water—that is, water containing hydrogen's heavy isotope deuterium rather than hydrogen-1. However, such reactors aren't used in the U.S.)

Sorting out isotopes of the same element is a formidable task. The oldest process for uranium enrichment is call *gaseous diffusion,* and it relies on the fact that at a constant temperature, lighter molecules and atoms move faster than heavier ones, with the difference in speed proportional to the difference in weight. Obviously, you need a gas for this process, and uranium is a solid, so gaseous diffusion uses uranium hexafluoride, a molecule containing uranium that becomes a gas at about 56° C (133° F).

Gaseous diffusion works. In fact, it's responsible for essentially all of the enriched uranium that exists today in the U.S. However, it's a very expensive process. Gaseous diffusion plants are huge, covering hundreds of acres—so big that employees drive cars and ride bicycles through the corridors. Each plant requires its own electrical power plant and consumes hundreds of millions of dollars worth of electricity each year. The price tag for the entire plant runs well into the billions of dollars. The process isn't even very good at collecting uranium-235; about one-third ends up in the "tailings" heap instead of in reactor fuel.

New technology is needed, and for the immediate future the leading candidate is the gas *centrifuge,* which also relies on the difference in weight between uranium isotopes, separating them by spinning a gaseous uranium compound. The DOE plans to build a pilot gas-centrifuge plant and begin operations within the decade. The plant would be expensive—well over $1 billion—but would use only 10 percent of the energy of a gaseous-diffusion plant with comparable output. Still, some scientists think we could do better.

With a laser of course. A laser can be used to energize a single isotope of an element only (or only molecules containing that isotope). The energized atoms (or molecules) of the desired isotope can then be made to react chemically with other substances so that they can be separated from the material that isn't wanted.

Laser enrichment of isotopes may be simple in theory, but it's complex in practice. One approach that's being pursued at the Lawrence Livermore National Laboratory, the atomic approach, involves uranium vapor. Unfortunately, vaporizing uranium means working at very high temperatures. As if the high temperatures themselves didn't present enough problems, uranium vapor turns out to be nasty, corrosive stuff—to say nothing of its being radioactive.

Many details of the process aren't being talked about in public, but the basic idea is to tune a laser emitting light in, or near, the visible region of the spectrum to the exact wavelength absorbed by uranium-235 and then somehow to collect the energized uranium-235, by either zapping it with another laser to ionize it or causing it to react chemically with another material.

A different laser approach, the molecular approach, is being pursued at the Los Alamos National Laboratory. There, researchers start with uranium hexafluoride gas, or perhaps another uranium compound. Molecules containing uranium-235 are then selectively energized with a laser tuned to a precise wavelength in the infrared region. A second laser emitting ultraviolet light illuminates the uranium hexafluoride, providing enough energy to free a fluorine atom from the energized molecules containing uranium-235. The resulting uranium pentafluoride molecules can be separated easily from the uranium hexafluoride molecules, because they have different chemical properties and liquefaction temperatures.

The atomic and molecular laser processes have two very important points in common. In theory, they consume very little energy per atom of uranium-235 in the finished product, because the laser energy goes only to excite the rare isotope. This advantage offsets the low efficiencies of some of the lasers involved.

An additional advantage is that both processes have great sensitivity.

Lasers are very efficient at picking out uranium-235. Laser enrichment plants should be able to reduce the uranium-235 content of the waste tailings to 0.05 to 0.08 percent—about 10 percent of the natural 0.7 percent level. In contrast, tailings from gaseous-diffusion plants contain about 0.25 to 0.30 percent of uranium-235. Indeed, one of the major reasons the Department of Energy is interested in laser enrichment is its potential to recover additional uranium-235 from the large existing stockpiles of tailings from gaseous-diffusion plants. In effect, laser enrichment would increase our uranium supplies by 30 to 40 percent.

Offsetting these advantages are some sizeable problems, however. The vapor process requires handling atomic uranium vapor, which, as we said earlier, is nasty stuff. The molecular process requires development of a suitable infrared laser, which does not yet exist. Both processes also require other lasers for the second step of freeing an atom from the excited molecule or an electron from the excited atom. And neither is quite as simple as we've made it sound.

The two laser processes are two of the three advanced isotope separation processes being studied by the Department of Energy. (The third is a laserless plasma process being developed at TRW, Inc.) The DOE hopes to pick one of these technologies for further development by mid-1982 and to put the other two on the back burner.

PROLIFERATION PROBLEMS

We haven't yet touched upon the most delicate—and most important—issue in laser enrichment of uranium: nuclear proliferation.

The problem is that uranium-enrichment technology has been the bottleneck that has helped limit the spread of nuclear weapons. It's not hard to mine uranium and refine it to get uranium oxide, or even pure uranium metal. Nor is it outrageously difficult to design an atomic (fission) bomb—a Princeton undergraduate student did it a few years ago. With present technology, however, it's very hard to enrich the uranium-235 content of natural uranium to the high levels required to build a bomb.

Both gaseous-diffusion and gas-centrifuge technology are difficult, complex, and expensive to implement. So far, both atomic-vapor and molecular laser isotope enrichment processes also appear to be difficult to implement—at least insofar as they require complex technology that isn't readily available.

Could an unexpected breakthrough in laser technology demolish this barrier and open the way to nuclear proliferation? It's possible but not likely. It would require both favorable physics and a fortunate—or unfortunate, depending on how you look at it—coincidence of laser output and uranium absorption at the same wavelength. It would probably be possible only for the molecular process, because there isn't any way to make the problem of handling atomic uranium vapor go away. It *is* conceptually possible, however, and that's a major reason why Los Alamos's program, which concentrates on molecular enrichment, is under particularly heavy security wraps.

Los Alamos and Livermore have produced macroscopic—readily visible to the human eye—quantities of enriched uranium. But the processes are not yet ready to be put to work. Years of engineering must come first, and neither laser approach is likely to begin enriching uranium on a large scale much before the end of the decade.

Meanwhile, isotope-enrichment researchers are working on applying similar laser technology to the purification of plutonium produced in nuclear reactors. This may have even more serious proliferation implications than uranium enrichment because the ultimate goal appears to be producing material suitable for use in bombs. In fission reactors, free neutrons convert some of the uranium-238 to plutonium-239, which, like uranium-235, can be split by neutrons in bombs or reactors. Enough other isotopes of plutonium are produced to make the resulting plutonium unsuitable for use in bombs (where purity requirements are more stringent than in reactors). DOE is working on ways to use lasers to remove the contaminating isotopes.

SOLAR-POWER-SATELLITES

One of the farthest-out energy concepts that's been proposed in recent years is the solar-power-satellite. Such a satellite would sit in geosynchronous orbit, which would make it appear stationary over one point on the earth's equator, collecting energy from the sun. That energy would then be beamed down to earth.

The leading contender for beaming the energy down to earth has been microwaves. While microwave technology is capable of doing the job—at least in theory—it presents its own set of problems. The biggest is the immense land requirement—about 200 sq km (75 sq miles)—for a microwave receiver, compared with about 1 sq km (0.4 sq mile) for a laser receiver, for collecting the same amount of power. Other concerns include the possible disruption of communication services, which are increasingly dependent on microwave transmission, and nagging uncertainties about the potential health hazards of microwaves. In the words of one laser advocate, "Do you want to turn the atmosphere into a microwave oven?"

Lasers have their own problems, however. Their higher power density would permit a smaller receiver but could also raise concerns about the effects of misdirecting a laser beam. Indeed, such a laser would have power levels higher than those required for laser weapons, although it would not need the elaborate beam-direction equipment required for laser weapons, because it would lock onto a "cooperative" target—the laser receiver on the ground. Nor would the beam have to be focused on as small a spot. Making certain that a power-transmission laser didn't have the capability of acting as a weapon, however, might present some interesting problems in arms-control verification.

At the moment, such concerns are somewhat academic. Advocates of space colonization and a greatly enhanced space program picked up the solar-power-satellite concept and pushed it so strongly that in 1979 it was labeled a "political football" by Donald Callahan, program manager for that part of the program operated by the National Aeronautics and Space Administration. The following year Congress punted the ball and deleted all funding for solar-power-satellite research from the fiscal 1981 budget.

11 THE LASER LEARNS HOW TO READ AND WRITE

Open up your cupboard, and you see striped symbols on food packages, designed to be read by a laser so that a supermarket clerk doesn't have to ring up prices on a cash register. Open your mail, and you may find a computer-generated statement, perhaps from a bank or insurance company, that was printed with a laser. Pick up one of many major newspapers, among them the *New York Times* and the *Los Angeles Times,* and the pages you see may have been printed from printing plates produced with a laser.

These are the uses of lasers in the information-handling business that touch us most directly, but they are only part of the picture. It all goes back to the computer, which has revolutionized the way we process information. No longer is information processing—analyzing data or performing calculations—the bottleneck. Instead, we are limited by the speed with which information can be put into and extracted from the computer: the input and output speeds, in the jargon of the computer world. So engineers are turning to lasers to read and write computer input and output in a variety of ways, ranging from deciphering those cryptic striped symbols on cans of soup to storing information on supercompact, computer-readable disks to be retrieved for later use.

For most of these applications, the compelling lure of the laser, from the engineer's standpoint, is that its beam can be focused to a spot that may be as small as 0.001 mm (0.00004 in) in diameter—much smaller than is possible with ordinary light sources. By scanning that tiny spot across a surface and monitoring the amount of light reflected, it's possible to read special symbols and even certain typefaces. If a laser beam is modulated in intensity as it's scanned across a light-sensitive material, it can write on the material at very high speeds. A pulsed laser can make tiny holes in a special recording medium that correspond to the "ones" and "zeroes" in which digital computers think.

Lasers do many demanding information-related jobs. Ultrahigh-resolution images from NASA's LANDSAT satellites are recorded with a special laser system. Laser-based optical memories hold out the promise of being able to store tens of billions of bits (fundamental units of information—each a zero or one) much less expensively than other kinds of memories. Optical computers can perform certain computations much more readily than conventional digital computers. And the much-heralded "office of the future" may include as one key component a combination laser printer and photocopier that can send, receive, and generate printed pages in computer-compatible form.

SCANNING THE UNIVERSAL PRODUCT CODE

The series of stripes on labels of products sold in supermarkets is called the Universal Product Code (UPC), and it is indeed fairly universal on prepackaged products sold in supermarkets across the United States. The bar code was adopted as a standard in mid-1973 by the Super Market Institute, to permit development of computer systems to automatically tally up purchases at the checkout counter without a clerk having to push buttons on a cash register.

The basic idea is fairly simple. Each separate, prepackaged product has a distinct code printed on the package. For example, each 10¾-ounce can of Campbell's tomato soup has printed on it a series of stripes corresponding to the code 51000–00011, while each 10¾-ounce can of Campbell's condensed chicken broth and vegetable soup has the code 51000–00117. Different-sized cans of the same soup would have different codes. The first five digits identify the maker of the product; the second five specify the product. When the code is read off the label, the computer looks up the price of the product in its memory. The price is entered into the computer by the store management and isn't necessarily the same as the price marked in human-readable dollars and cents on the package (although in practice differences between the two prices generally represent mistakes by the store). The computer automatically tallies up all purchases and rings up a total.

One obvious thing this does for a store is to speed up checkout. If an

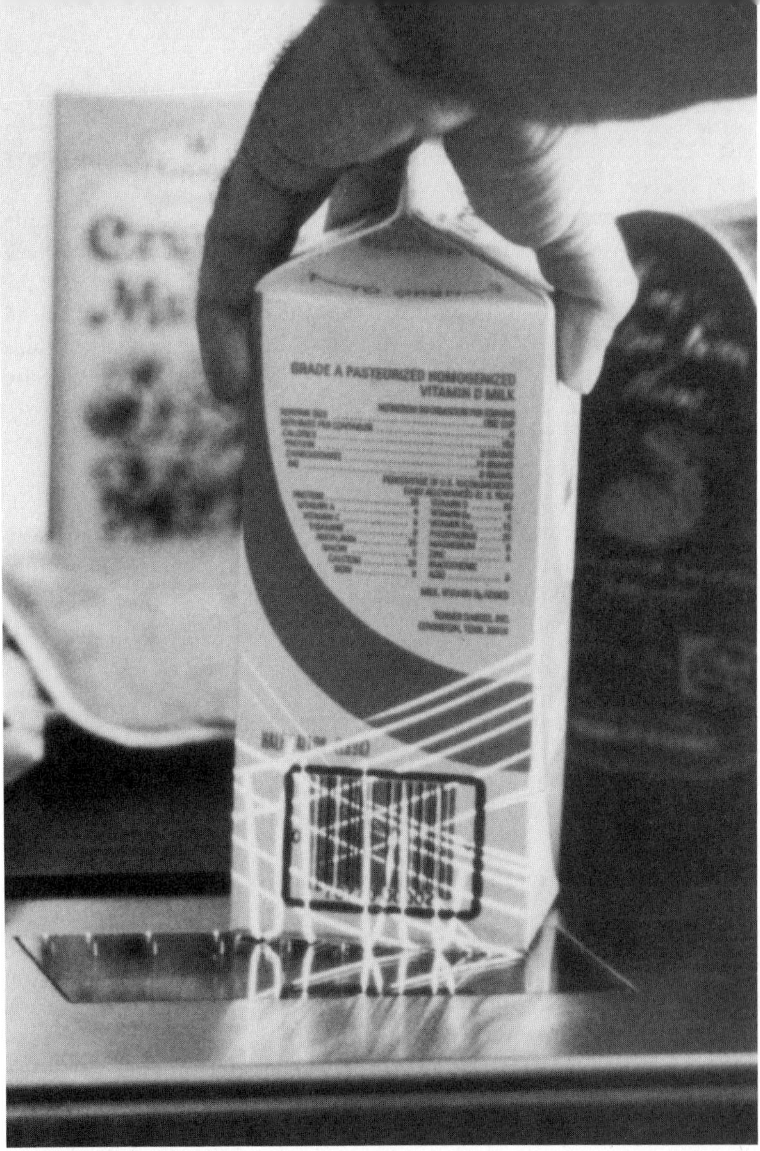

A holographic laser scanner wraps its beam around the UPC label on a half-gallon carton of milk. The price is automatically rung up, and one half gallon of milk is subtracted from the inventory list. Courtesy International Business Machines Corporation

item has a UPC label, it can be passed over a laser scanner that reads the label, then packed right into a bag, without the clerk having to punch keys on a cash register. Another big, but not so immediately obvious, advantage is inventory control. The computer can automatically tally up how many cans of tomato soup are bought and automatically order more when the supply starts getting low. (It is necessary to check periodically how well the computer's records conform to what's on the shelves—particularly if shoplifting is a problem—but such checks need not be as frequent as ordinary inventories.) Laser scanning also avoids mistakes by harried clerks, which the supermarket industry insists are more often than not in the customer's favor. Finally, in theory it eliminates the need to mark a price on each item—although that's an advantage that few stores have been able to make use of, for reasons we'll get into later.

The striped format of the Universal Product Code reflects many technological factors. For "universal" use of the Universal Product Code to be economical, the code has to be printed along with the rest of the package label. Thus the UPC symbol has to be printable with ordinary printing presses. The most critical factor in making sure labels are read correctly turns out to be the spreading of ink during printing, which is much easier to control perpendicular to the direction in which paper runs through printing presses than it is parallel to that direction— hence the choice of a striped code, which is aligned so that the critical dimensions are perpendicular to the direction in which presses roll.

The nature of a laser beam, with its sharp focus and small size, makes it particularly well-suited for reading the bar-code symbol. Helium-neon gas lasers, which emit red beams, were chosen for this task, largely because they are dependable and are the least expensive type of laser with visible output (important because printing must be checked visually).

The laser reader also contains some means of scanning the laser beam very rapidly through a fixed geometric pattern (different makers of laser scanners use different means and different patterns). The reflected light is monitored by an optical detector inside the reader. When the reader sees the unique pattern of reflected light that corresponds to a valid code, it transmits the code to the computer, which looks up the corresponding price and adds it to the running total it keeps for each customer. The price is also printed out at an automatic cash register at the checkout counter, along with a descriptive label (e.g., CAMP TOM SOUP 0.25). The scanner also beeps to indicate that it's read the symbol.

The laser itself, the optics that scan the beam, and the associated electronics are mounted below the counter. The beam emerges through a piece of glass or plastic or an open slot. The clerk at the checkout counter passes each item over the scanner with the UPC symbol down so that the scanner can read it; if the scanner doesn't beep, the clerk does it again. If an item doesn't have a UPC label (for example, a bunch of bananas), the clerk pushes keys on the cash register. You can see the pattern scanned by the laser by holding a piece of paper or other object close to the scanning window. The beam power is low enough, and the beam scanning fast enough, that accidentally looking at the beam can't do any harm to your eyes.

There's even a talking scanner, which was developed by the National Semiconductor Corporation of Santa Clara, California. It includes speech-synthesis circuits, so that its artificial voice can tell you how much you're being charged. The scanners can even be programmed to say, "Thank you for shopping with us today." A National Semiconductor press release says the talking scanner "will bring back an old friend to the counter," but we're not convinced.

The technology for automated checkout was ready by about 1975, but scanner-equipped supermarkets were a novelty until about 1980. Part of the delay was due to a natural hesitation on the part of supermarkets to adopt a new technology. But a large part was due to customer-relations blunders on the part of the supermarket industry.

Scanners became available shortly after a severe round of food-price inflation, when it seemed as if stores raised all their prices every week. At the time, the most obvious advantage of scanning was that stores could avoid marking prices on each package. Although stores pointed to unit prices posted on the shelves, few customers were ready to trust them and rely on shelf labels—largely because many stores use cryptic computer-generated labels and are often careless in matching shelf labels with goods. Most stores quickly sensed consumer hostility to unmarked prices and quickly backed off from the idea. Some states now explicitly require that prices be marked on all packages in "human-readable" format. Only a handful of warehouse, or discount, stores have stopped marking prices on each product.

Scanners turn out to be well worthwhile even for stores that mark prices on every item, but it took the supermarket industry a while to realize this. Tens of thousands of scanners are now in use in thousands of supermarkets across the United States. A few are at work in Canada. Japan and Western Europe have developed their own UPC-compatible codes.

Laser-read bar codes similar to the Universal Product Code have found many other applications, such as helping blood banks keep control of their inventories. Outside of the supermarket, you're most likely to have seen UPC-like symbols on the covers of books and consumer magazines distributed through newsstands. They appear at the lower left corner of most magazine covers and are intended for use by wholesalers, who count the covers of unsold magazines (only the covers are returned to the wholesaler; the rest of the magazine is supposed to be discarded). The torn-off covers zip by a laser scanner, which keeps count.

What's a boon to wholesalers is a nuisance to the art directors who are responsible for creating attractive magazine covers. In its typical way, *Mad* magazine made its displeasure obvious; on the cover of one issue a small figure attacked the bar code with a lawnmower, which spewed out fragments of bars in its exhaust.

Omni, a science magazine that has won several awards for its artwork and design, refused for almost two years to print UPC symbols on its covers. The reason? As an *Omni* designer explained: "They're just ugly—really disgusting." Even *Omni* had to give in though, and in mid-1980 that familiar, but "disgusting," pattern of stripes appeared on the proud magazine's cover. The final blow, according to Robert Castardi in *Omni*'s circulation department, came not from the wholesalers but from supermarket chains that refused to display the magazine in their racks. Safeway stores, among others, wanted their clerks to be able to laser-scan *Omni* along with every other product. In retrospect, it seems ironical that it was a magazine devoted to futuristic science and technology that put up one of the final struggles against the laser's use in retailing.

Probably the most unusual application of laser scanning of bar codes is in reading ski-lift passes. The first such system went into service in the winter of 1979–80 at the French Alps ski resort of Puy Saint Vincent. A laser scanner at the turnstile automatically reads a bar code on a 13-by-8-cm (5-by-3-in) plastic card. Each skier has a unique code, which indicates how long he is authorized to use the lifts. The system, developed by Option S. A. of Meylan, France, is intended to speed the flow of skiers through turnstiles and keep out unauthorized persons, while reducing the number of attendants required.

READING MANUSCRIPTS

Lasers can read letters and numbers as well as specially designed symbols. The idea is the same: the laser beam is focused to a fine spot and scanned across whatever is to be read, and the reflected light is monitored and analyzed. The ultimate goal is automatic reading of the same typefaces that humans read, but at this point it's still a long way off. Only certain, specially formatted characters can be read by a laser; and the laser is only one of several approaches to what's called *optical character recognition.*

The promise is real, but the achievements are modest so far. One of us worked for a few years with a laser system that could read characters typed with a special ball on an IBM Selectric typewriter using a special ribbon. It was a fairly common system in the printing industry but a frustrating one. Manuscripts had to be typed in a special format that didn't allow much room for corrections, and since few writers can type a manuscript flawlessly as they write it, this meant an additional person was needed to retype the manuscript. Marks made with anything but a red or pink felt pen (both of which are invisible to the red laser beam) could make the manuscript unreadable to the laser system. The reading was also far from flawless. Mistakes were common enough that there was no escape from proofreading everything both before the retyped manuscripts went to the printer and after the typeset material came back. Finally, the system would sometimes get upset over flaws in a typewriter ribbon or other factors too subtle for the human eye to detect and produce typeset material full of mistakes. Developments in printing technology are fast enough that this system, introduced in the mid-1970s, is now well on its way to obsolescence.

MAKING PRINTING PLATES

The new generation of printing technologies includes laser technology that can do far more than read text typed in a special format. Laser systems that can produce printing plates are already being used to publish newspapers at plants far from the editorial offices. The same technology will also make it possible to transfer information stored in a computer directly onto printing plates—without any intermediate typesetting. To understand the significance of these capabilities, let's first stop to see how a newspaper works now.

Most of this country's newspapers have retired their linotype machines—clattering mechanical contraptions that cast slugs of type from molten lead. Instead, they set type using a process that produces a master photographic copy of the typeset material. This "cold" type (in essence a photographic print) is eventually pasted onto an artboard in the configuration that's desired for the printed page. This paste-up is then photographed, and the negative is used to expose a printing plate, which is then mounted on a printing press and used to print a newspaper (or magazine or book).

This process is gradually becoming outdated by the advent of computerized typesetting, which is particularly common on newspapers, where turnaround times and deadlines are most critical. In this case, writers and editors work at computer terminals, entering material directly into the computer's memory. Once a story is completed, the text can be turned into finished cold type on demand, without any intermediate stages or proofreading.

The first practical laser systems for producing printing plates actually used two lasers. The first was a low-power helium-neon laser that read the completed, pasted-up page into the memory of the computer system. The computer then modulated the intensity of a second, higher-power laser (argon or neodymium-YAG) as the beam was scanned across a specially prepared printing plate. The laser thus produced photographically an exact copy of the material on the master pasted-up page on the printing plate.

Such laser systems are in use at dozens of large newspapers around the world. They are particularly advantageous for newspapers that are printed at two or more locations, such as the *New York Times* or the *Los Angeles Times.* Information read off a master, pasted-up page at the main office can be transmitted simultaneously to remote laser writing systems at two or more locations. Such printing arrangements are commoner in other countries than in the United States, and as a result, there is a heavy concentration of such laser systems in Europe, where many newspapers are distributed nationally.

The ability to produce printing plates at remote printing plants is by no means the only advantage of the laser platemakers. One of the earliest users, the *Las Vegas Review-Journal,* reported in mid-1977 that the $200,000 laser system it purchased a year earlier from the Eocom Corporation of Tustin, California, saved about $100,000 per year, half in labor and half in materials. Another advantage is the ability to expose a printing plate in only four minutes, compared with half an hour for conventional hardware. According to Louis Harga, production manager of the Las Vegas paper, the laser system produces plates of the same quality as the previous system, is more tolerant of dirt, and ignores the edges of scanned materials, which can sometimes show up as unwanted lines on the printed page.

The next step is to produce the printing plate directly from the computer without having to make and scan a master, pasted-up page. It's been demonstrated by Eocom but has yet to come into widespread use. In this case, the computer feeds the information needed to generate the final type directly into the laser system. The laser reading unit records photographic images, line drawings, and advertisements submitted in finished form to the newspaper, and converts them into computer-compatible form. An editor working with the computer arranges the material. Then the computer uses the laser platemaker to expose the printing plate.

The laser-based computer-to-plate system would save many time-consuming intermediate steps and thereby save money for newspaper publishers. Those savings would come at the cost of work for newspaper employees, and hence the idea is coming under fire from unions. Indeed, one cause of an 88-day strike at the *New York Times* in 1978 was the paper's plans to install new equipment, including several laser platemakers from LogEScan Systems of Springfield, Virginia—a battle that the *Times* management won.

FACSIMILE TRANSMISSION

Producing printing plates that replicate pasted-up originals is really a special case of the general problem of facsimile transmission—producing a copy of an original document at a remote location. Facsimile transmission is a technology that's never really gotten very far off the ground, but it does serve many specialized functions, such as transmitting wire-service photos and weather maps. A small fraction of these needs are met with laser-based equipment.

Associated Press photographs used by newspapers are produced on

"Laserphoto" receivers developed by the Harris Corporation. These recorders use a helium-neon laser beam, which is modulated in intensity and focused to a fine spot. The laser writes on a special photographic paper that is developed not by a wet chemical process but simply by heating it. The process is called "dry silver" (because, like ordinary photographic papers, it uses silver but, unlike them, is processed without "wet" chemical solutions). This process is suitable for the high quality required for photographic reproduction but not for general-purpose facsimile transmission because of the high cost of the silver-containing paper.

Another approach was taken by the Xerox Corporation, which combined laser scanning with photocopier technology. Xerox photocopiers, as well as those of most other manufacturers, work by forming an image of the original document on a drum coated with a light-sensitive material. The image recorded on this drum is then transferred to a piece of paper. The Xerox laser facsimile system (called Telecopier 200) receives images by writing directly on the image-transfer drum with a helium-neon laser; that image is then transferred to paper. To transmit an image, the same laser is used to scan an original document, and measurements of the reflected light are converted into a format suitable for transmission to another facsimile receiver. A major advantage of the laser system is its speed. It takes just two minutes to send an image, only half as long as a laserless facsimile unit offered by Xerox.

THE OFFICE OF THE FUTURE

Similar laser technology may come to play a much more important role in future office equipment. To understand how, let's take a brief look at a concept described by a much-overused term: "the office of the future."

Much of the information processed in modern offices is in the form of words, and much of that processing can be computerized. Specialized computers called word processors are on the scene; despite their name, they *are* computers, but computers designed to handle words rather than numbers. The next logical step is the development of word processors that can communicate directly with one another without human intervention. A few such systems are already on the market, including models priced at $50,000 to $100,000 from two industrial

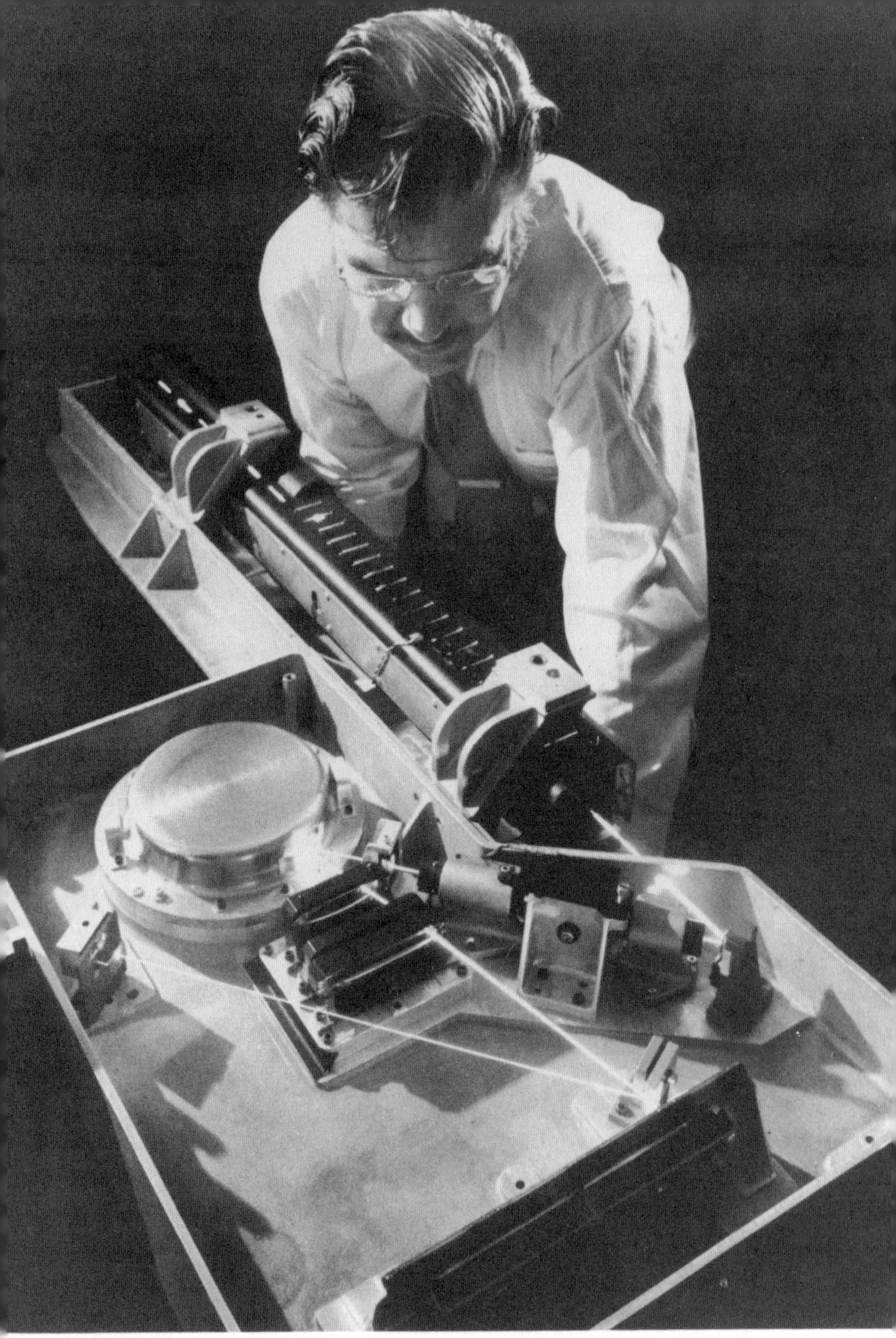

A high-speed IBM printing system is uncovered to reveal the helium-neon laser inside. The low-power laser forms characters on a rotating drum. As the drum turns, a dry, inklike powder sticks to the images written by the laser and then is transferred to paper. Courtesy International Business Machines Corporation "heavy hitters," the Xerox Corporation and the International Business Machines (IBM) Corporation.

The printing ends of the Xerox and IBM systems are strikingly similar: each uses a laser to write on a photocopier-type drum. The printer can serve as a facsimile transceiver (a transmitter/receiver), or it can, in some versions, simply function as a photocopier. It can also, more importantly, serve as an all-purpose printer, generating entire business forms, signatures, and company logotypes, as well as numbers and letters.

The scanning technology is similar to that used to expose printing plates, but the resolution is generally somewhat less, because photocopying technology can't achieve as fine a resolution as most printing (something which becomes painfully obvious after looking at a document that has gone through several generations of photocopying).

While a laser facsimile printer produces one page every two minutes, the Xerox model 5700 Electronic Printing System (a communicating word processor with a number of other "bells and whistles") can transmit or receive a page in only three seconds. When operating as a printer, it can produce up to 43 pages per minute. Even higher speeds are possible in specially designed laser printers, some of which can write up to about 300 pages per minute.

So far, it's these extremely high-speed laser printers that have been the most successful in the marketplace. The reason is their speed and versatility. Computers can generate vast quantities of information, and in the past a single large computer often kept several output printers busy at the same time. A high-speed laser printer can churn out forms or whatever at several times the rate of ordinary "impact" printers (so-called because, as in a typewriter, a key strikes a ribbon that writes on the paper). What's more, the laser printers produce uniform impressions and can be programmed to have much more attractive typefaces than the capital letters that are all that are offered by most impact printers. The result looks like a good photocopy of a printed original; when you hold it under a magnifying glass, there's no sign that the type was produced by scanning with a laser beam.

The systems, at some $200,000 or so each, are expensive. However, their speed and versatility can be invaluable to organizations that have to produce a large volume of computer output—whether in the form of individualized mailings to customers (such as money-market-fund statements or personalized junk mail) or internal reports.

SATELLITE PHOTOS

Lasers have also been put to work recording high-resolution images. The National Aeronautics and Space Administration (NASA) uses laser systems to record images transmitted from space. Those striking LANDSAT images you've probably seen, for example, are produced from signals transmitted from a satellite. On the ground, NASA uses a laser system built by the RCA Corporation to record the photographs on film for analysis by ground personnel. The resolution is exceptional—the laser system can write as many as 20,000 separate spots on a single line on 24.1-centimeter-wide (about 10 in) film. A similar RCA-built laser system records data from NASA's synchronous weather satellite.

OPTICAL MEMORIES

In chapter 13, we'll talk about how lasers are being used in videodisk players, which play back prerecorded video programs. Whatever you think of the content of television programs, each one represents a tremendous amount of information. Some videodisk developers turned to laser technology to store and retrieve that information. Technology similar to that used in the laser-equipped home videodisk player developed by N. V. Philips and MCA, Inc. can also be used to store digital data for computers. Philips, RCA, and others have demonstrated impressive capacities in prototype systems, and commercial systems are clearly on the way, although they are not on the market as of this writing. Developers talk of storing 10 billion bits or more on a single side of an optical disk. How is all this possible?

The key point, once again, is the fact that a laser beam can be focused down to an extremely small spot. A laser can record information in the form of holes only about one micrometer (one-millionth of a meter, or 0.00004 in) in diameter, burnt into layers of light-sensitive material. This material is generally a thin film, which is coated onto a disk, which revolves very rapidly. Extremely short laser pulses are used to record the information, so that the spot does not spread out too much as the disk turns.

The information thus recorded with a laser is also read back with a laser, although at a lower power, so that no holes will accidentally be produced. Once again the laser is focused to an extremely small spot, and the disk rotates, but this time the light reflected by the disk is monitored to extract the information recorded earlier.

One thing you can't do with present optical memories is erase data and write new data in its place, something that's possible with existing magnetic data storage equipment. The inability to write over old data makes it impossible to update information extensively without creating a whole new optical disk. A modest amount of correction is possible on an optical disk by writing on areas intentionally left blank, but for many applications, the ability to erase old material and substitute new information in any desired quantity is essential.

However, there are situations in which it's possible to live with the inability to erase because of other advantages. It should cost much less to store a bit of information on an optical disk than on conventional magnetic disks or tapes—and it looks like the information will stay in usable form much longer, too. Optical storage also offers tremendous capacity without the usual penalty of slow access to stored information.

The first applications of optical disks are likely to be in jobs where conventional computer storage isn't economical. One example is the extensive file of documents on land titles compiled by Ticor, a title insurance company based in Los Angeles. The company plans to record digital images of the documents and store them in an optical memory (the model they've picked uses slides rather than disks, but the idea is the same). Any time information on one of the documents is needed, an image of the document can automatically be retrieved from the optical memory within 12 seconds—much faster than a piece of paper could be found in a bank of file cabinets.

Optical memory technology may find other applications outside of disk formats. Jerry Drexler, president of the Drexler Technology Corporation in Mountain View, California, has proposed using a stripe of optical recording material on the back of credit cards, rather than the magnetic strips now in use. The large capacity of the optical material would provide plenty of room for recording and updating the cardholder's credit rating and other relevant information. That proposal is now under consideration, along with a competing idea to include a microcomputer chip in every credit card, an idea that Drexler says would be much more expensive to implement. Drexler has also suggested using optical cards as portable personal medical histories.

OPTICAL COMPUTING

So far we've talked about techniques of laser reading and writing that are hard at work, or nearly so, in the real world. The laser equivalent of arithmetic—generally called optical computing for reasons we'll get into shortly—has found few applications outside the laboratory to date. The topic is a complex one, and we can't hope to do it justice here, but it offers some tantalizing promises.

It's possible to build optical equivalents of the digital logic functions found inside a computer, although it's difficult. Some researchers are trying to develop "integrated optics," in which several optical devices would be fabricated in a thin film on a block of another material. Someday, such integrated optical circuits may perform some of the same functions as today's integrated electrical circuits. But at the present time the major allure of optical computing seems to be the ease with which it can perform some operations that are relatively difficult for ordinary electronic computers.

The basic idea is that a lens can perform a mathematical function called a *Fourier transform* on a pattern of light (perhaps from a laser) passing through it. This isn't your ordinary arithmetic—it's something you wouldn't encounter until an advanced calculus class. But it turns out to be useful. For example, suppose you have a satellite photograph that has a series of parallel lines across its surface, and you want to get rid of the lines. When you take a Fourier transform of that image, the light from the lines is separated from the other light in the image. That makes it possible to block the light from the lines before you take another Fourier transform with an identical lens to get the original image back, minus most of the lines. Neat!

A more complicated version of the same principle can be used to recognize an image automatically. Let's say you're a general in the army presented with a large stack of aerial photographs. You're looking for an enemy tank, but you don't want to spend all week examining each photograph in detail. So, very roughly speaking, you superimpose a Fourier transform of the tank on a Fourier transform of each photograph. Then you perform another Fourier transform on the superimposed Fourier-transform images. Superimposing the Fourier transform of the tank on the Fourier transform of the photograph essentially filters out everything in the image but the tank. So when you look at the second Fourier transform, the image of the tank should jump out at you if it's present, because what you're looking at is a reconstruction of the photograph with everything but the tank removed.

The military has other potential uses for such techniques. They'd like to develop weapons that can automatically recognize, home in on, and destroy the enemy. This gets tricky, because you must first make very fine, but critical, distinctions, such as telling one of our tanks from one of theirs.

There's still a long way to go before image recognition techniques become practical. However, optical processing is being used for image enhancement in a special type of radar system called *synthetic-aperture radar,* which attains high resolution despite its use of small antennas. Optical computing techniques have found other uses in image enhancement, many of them limited to the laboratory so far.

Optical computing doesn't inherently require a laser, although, in many cases, using a laser's coherent, monochromatic output is a big advantage. In other cases, the coherence introduces undesired noise. Exactly what the laser's role will be in the long term is unclear, because the technology is still in its infancy.

12 HOLOGRAPHY: A KIND OF MAGIC

There's something both magical and elusive about holography that you can't really understand until you meet a hologram face to face. When the lighting is right, it creates an unmistakably three-dimensional image in midair. Still, you would never mistake the holographic image for the real object: the colors are not natural, there's a subtle graininess, and the image fades or disappears if you view it from the wrong angle. Even so, your instincts tell you to reach out and touch it; but when you close your hand, it will touch only thin air.

Next time you're in New York City, pay a visit to the Museum of Holography at 11 Mercer Street in Manhattan's colorful Soho district. It's tiny as museums go, with ultramodern technology housed in a 110-year-old building on a decrepit side street. Inside you'll find a small group of people excited about the potential of holography, who can point you to other people working in other places who feel the same way. The community is tiny, but it is a community.

Spend some time in the museum, and you can see both the strengths and weaknesses of holography. You'll see some holograms that are primitive technically and some that are primitive artistically. You'll appreciate how far holographers are from fulfilling the science-fiction scenarios of room-filling three-dimensional images. But you'll see how artists are learning technically and artistically to extract the most from a new medium.

Along with the promise come problems. The potential of holography has been oversold time and again, and many parts of the laser community have been burned. Indeed, there's a common, and not completely unjustified, cynicism that is best summed up by the discouraged engineer who complained that "the only people making money from holography are the pornographers." (More about that later.)

In this chapter, we'll take you on a tour of the fascinating world of holography. We'll stop to deflate some false expectations and to point out some wonders that you've never heard about. We'll talk about some applications most people are unaware of; for holography, besides being an art form in and of itself, is also being used in video games, for testing industrial products such as airplane tires, and even as an aid in restoring paintings and sculptures that are centuries old.

First, though, we have to tell you what holography is, and to do that, we'll have to go back and take another, brief look at the physics of light.

THE HOLOGRAPHY CONCEPT

Holography is much more than a technique for producing three-dimensional images with a laser. It's a whole concept.

To understand it, we must look at light as electromagnetic waves. Earlier we concentrated on one important property of electromagnetic waves: their frequency. Now we need to look at two others, which we mentioned only briefly: the amplitude of a wave (how big it is, measured from peak to valley) and the phase (where it is in its cycle of oscillation from peak to valley and back).

The light we see is actually light reflected by the objects about us. From a physical standpoint, what these objects do is create a *wavefront* of reflected light. The wavefronts *contain all the information that reaches your eyes.* Your eyes and brain decipher this information so that you can see. Let's suppose you could record precisely a wavefront in a plane in front of you—a big horizontal slice of light in front of your eyes—and that you could later re-create the exact same wavefront. In theory, the re-created wavefront would show you the exact same scene as the original wavefront. The image from the re-created, or reconstructed, wavefront would appear to be every bit as three-dimensional as the real scene. The basic idea of holography is to re-create that wavefront precisely.

Now you may be saying to yourself: Isn't that what a camera does? Doesn't it capture a wavefront, a plane of electromagnetic radiation, on a flat piece of film? In a word, no. As we mentioned earlier, there are two essential quantities involved in reconstructing wavefronts: amplitude and phase. Photographic film and electronic detectors see only intensity, which depends on amplitude. They're insensitive to phase. That's why they produce images that are only two-dimensional.

In fact, your eyes also see only amplitude. However, your two eyes see the world from slightly different angles, and your brain automatically compares the subtly different two-dimensional images from each eye to generate a three-dimensional view of the world. The subtle differences in the way the world looks from different angles are due in part to the invisible phase information contained in the wavefront, and in effect your brain is automatically reconstructing the phase information when it synthesizes a three-dimensional image.

The same principle is used in photographic tricks that create three-dimensional images. Two cameras take pictures from slightly different angles, then the pictures from one camera are shown to one eye, and the pictures from the other camera are shown to the other eye. Each eye then sees a different two-dimensional image, and the brain automatically reconstructs a three-dimensional image, just as it would if the eyes were seeing the scene itself, instead of photographs of it.

Holography is the technique of recording information about both amplitude and phase. It was the brainchild of the late Dennis Gabor, who conceived the idea in 1947 and first demonstrated it in 1948. While the first demonstration was with light, Gabor's real goal was to improve the resolution of the electron microscope. (More about that later.)

Gabor's idea was simple, and it earned him the 1971 Nobel prize in physics. He superimposed two wavefronts of light. One was from the object being illuminated; the other was a reference beam—a beam that bypassed the object. Gabor then recorded on a photographic plate the intensity pattern where the two wavefronts were superimposed. Diagram 12 shows how this is done in a typical holographic setup. Because the reference beam was controlled precisely and could easily be recreated, it could later be shone through the photographic plate to reconstruct the wavefront produced by the object.

What is recorded on the photographic plate depends on the way in which the two sets of light waves—one from the object and one from the reference beam—add together, a phenomenon called interference.

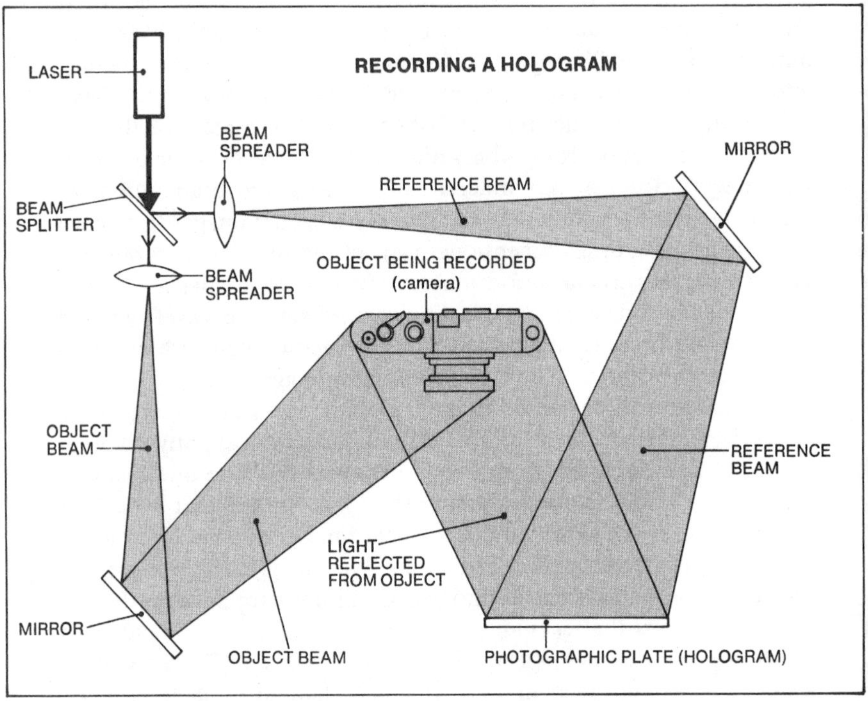

12. To record a hologram, the beam from a laser is split into two beams: an object beam and a reference beam. Both beams must travel about the same distance to the photographic plate, but the object beam is reflected by the object to be recorded. When the two beams come together at the photographic plate, they form an interference pattern, which doesn't look like an image at all. A three-dimensional image of the illuminated object is re-created when another laser beam, identical to the original reference beam, is passed through the interference pattern recorded on the photographic plate. (We have used a camera here as the object being recorded, because a camera is a common object that can easily be recognized from different angles.) Some types of holograms can even be reconstructed without a laser.

As we explained in chapter 9, this means that the instantaneous amplitudes of the waves are added together when superimposed. If one wave is at a peak and the other at a valley, the result is darkness; if both are at a peak or both at a valley, the result is light. This happens because what you see is intensity, which is the square of amplitude, which in turn is the distance between peak and valley of the combined wave. Cycle-to-cycle variations are much too fast to be seen by the human eye, or even the fastest electronic detector.

The hologram itself is a pattern of dark and light areas that represents the interference pattern produced by superposition of the two wavefronts. And it is these dark-and-light patterns that embody that second critical property needed for a three-dimensional image: phase. Where did phase come from? Not from magic but from the reference beam. Remember that the dark-and-light patterns of the hologram were produced by interference, which depends on the phase of both the waves involved. If you know the interference pattern that was produced by the light from the object and the reference beam, and if you can precisely re-create the reference beam, in principle you have all that you need to re-create the wavefront from the object.

The hologram bears no obvious resemblance to the object, and in fact, most holograms look like random patterns. But a hologram allows you to re-create the recorded object whenever you want to, simply by illuminating the photographic plate containing the hologram with the reference beam. *Diffraction,* the special way in which the reference beam spreads out as it passes through the interference pattern (the hologram), re-creates the wavefront from the object, which then travels to the eye as if it had never been interrupted by the photographic plate.

By the way, the same image is produced whether the hologram is a negative or a positive print. "It turned out that in holography Nature is on the inventor's side," Gabor said in his Nobel lecture.

There's one requirement for holography we haven't discussed yet: coherence. When we talked about the basics of lasers, we described coherence as waves marching along exactly in phase. Holography only works if the light from the reference beam and the light coming from the object are marching along exactly in phase with each other. Coherent light must also be monochromatic—all of the same wavelength.

What this means in practice is that the light for both beams must come from the same source. The distance the two beams travel can't differ by more than the coherence length of light—that is, the distance over which the light waves stay precisely in phase. The coherence length of an ordinary light bulb is essentially zero. Coherence lengths for lasers are typically several meters (or yards), sometimes longer.

Gabor worked on the holography concept for a dozen years before the first laser was built, and the best light source he could find was a high-pressure mercury lamp, which had a coherence length of only 0.1 mm (0.004 in), or roughly 200 wavelengths of light. This put severe restraints on his experiments. To make sure that both beams traveled the same distance, he used an optical design in which the reference beam itself illuminated the object, which was a transparent photograph only 1 mm (0.04 in) in diameter. The hologram itself was only 1 cm (0.4 in) across, and it still took several minutes to record it on the most sensitive film available. Gabor's technique was also plagued by stray scattered light obscuring the hologram and the reconstructed image, and his optical arrangement didn't separate the reconstructed image from other light sufficiently. After an initial flurry of research, interest died down, and by the mid-1950s, the field seemed to have gone into hibernation.

LEITH AND UPATNIEKS

Despite the apparent quiet, things were astir in holography. At the University of Michigan's Willow Run Laboratory in Ann Arbor (now the Environmental Research Institute of Michigan), Emmett N. Leith and several colleagues were working on holographic theory. They published nothing at the time, however, because they were applying their ideas to a new type of radar system, which was then classified.

Leith worked quietly from 1955 to 1962 and has since said that he had "quite phenomenal success" using mercury lamps as sources during that period. His theoretical work began paying off at the same time that lasers became readily available. Leith and Juris Upatnieks developed a new optical arrangement, in which light was split into a reference beam and a separate beam that illuminated the object and was then superimposed upon the reference beam to form a hologram.

Leith and Upatnieks used both lasers and mercury arc lamps in their early experiments. "Each [light source] had its own special advantages, and it was not entirely clear which was better," according to Leith. "We eventually chose the laser, but we found that good-quality holograms could be made with either source; the decision was a matter of which set of advantages to exploit and which set of problems to attack."

While credit for the idea of holography (along with the Nobel prize) went to Gabor, it was Leith and Upatnieks who brought it into the real world. Their first laser experiments were with flat objects, but they soon moved on to three-dimensional objects. In Leith's words, "This involved little new theory, but considerable new experimental technique."

Leith's comment is something of an understatement. Making holograms of flat objects didn't require much in the way of coherence, but working with three-dimensional objects did. A laser became an absolute necessity, because the coherence length required for holograms of three-dimensional objects is about twice the depth of those objects.

It also became essential to damp down all possible vibrations, such as those caused by a passing truck or a furnace turning on or off, because they were found to blur the reconstructed image. Damping out such vibrations requires a massive optical table, which must be isolated from vibrations transmitted by the floor. The first holographers used heavy sheets of metal on inner tubes from trucks. Modern laboratories use specially designed tables that can cost thousands of dollars.

MAKING THE HOLOGRAM

Let's stop to look at how you make a hologram of a three-dimensional object by the Leith-Upatnieks method.

You start with a single laser that emits a single beam. You then split the beam into two separate beams, using a special mirror called a beam-splitter, which transmits part of the light and reflects the rest. One part of the beam illuminates the object in such a way that the reflected light ultimately strikes a photographic plate or film. The rest is directed at the same plate or film, although it travels a different path. The lengths of the paths must be approximately the same. The superposition of the two beams exposes the photographic plate or film to produce an interference pattern. The film or plate, when developed, becomes the hologram.

In general, special film is used for holography, because much higher resolution is needed than for ordinary photography. Holographic films are specially sensitized to red light, because holograms are generally taken with helium-neon lasers, and ordinary films are not very sensitive to that laser's wavelength.

The hologram, with its interference pattern, bears no obvious resemblance to the object from which it was made. In Gabor's words: "It looks like noise."

But it isn't. You can reconstruct an image of the object by illuminating the hologram with light that is like the reference beam of laser light that went directly to the holographic plate. The hologram diffracts the light in a way that replicates the wavefront of the light reflected from the object. It's as if the wavefront had been frozen in the hologram until turning on the reconstructing beam released it to continue on its path to your eye. Your eyes see the same wavefront they would have seen if the light had been reflected from the object and are fooled into thinking they see a three-dimensional image.

There are, of course, real constraints on what holography can do. In general, the larger the object, the harder it is to make a hologram of it. This is because the illuminating light must be coherent over something like twice the depth of the object, and also because of practical considerations, such as the difficulty of making big enough holographic films and plates. It's pretty hard to record an image of an object over a meter (or yard) long, although it has been done.

Holograms also look grainy. This graininess is called *speckle,* and it's an inevitable consequence of irregular variations in the phase of coherent light caused by interactions with the atmosphere, reflective surfaces, and just about anything else that the light encounters. "It is not really noise; it is information which we do not want," in Gabor's words. He was referring to information on tiny irregularities of surfaces and fluctuations in the atmosphere. Speckle effects can be put to constructive use, but often they're a problem.

So far we've only talked about monochromatic holograms—holograms that are all one color, the color of the illuminating beam. Lifelike color reproduction continues to elude holographers, though there has been progress, which we'll get to shortly.

Holograms have some unusual properties that seem to defy common sense. For example, in most cases you need only part of a hologram to reconstruct the entire image. That's because the hologram records the same wavefront across its entire surface, and the pattern of that wavefront is, in essence, repeated many times. Thus, illuminating part of the hologram is enough to reconstruct the entire wavefront, although in practice there is a slight decline in quality when only small portions are used.

You can also record multiple holograms on the same photographic plate using different reference beams, or by directing the reference beam at different angles. If the holograms are recorded properly, a viewer will see only one reconstructed image at a time, even though several are recorded on the film.

WHITE-LIGHT HOLOGRAMS

What we've described so far is the relatively straightforward type of hologram initially developed by Leith and Upatnieks. You make the hologram with a laser, and you re-create the image with a laser. However, other types of holograms have been developed that don't require a laser to reconstruct the image. They still need a laser to make the hologram, though.

While Leith and Upatnieks were working on their holographic technique in the early 1960s, Yu. N. Denisyuk was taking a different approach in the Soviet Union. Earlier images had been re-created by shining a light *through* the hologram. Denisyuk developed a theory of *reflective,* rather than transmissive, holograms.

Denisyuk figured out that you could reflect the image off the hologram, and that you could illuminate it with white light, rather than a laser. By white light we mean ordinary light, such as you get from a light bulb or a fluorescent tube. While white light is a mishmash of different wavelengths, within that mishmash is light with a wavelength that matches that of the original reference laser beam. And that light can reconstruct an image from a reflection hologram, while light of other wavelengths is, in effect, ignored. The image itself is all one color, even though the illuminating light isn't.

Denisyuk developed his ideas in 1962, but, lacking a laser, he couldn't really demonstrate his technique. The first such reflecting holograms were instead produced by Americans George W. Stroke and A. Labeyrie in 1965.

Reflection holograms hold out the promise of full-color holography. The idea is to illuminate the object with three or more different lasers, in a variety of colors sufficient to reproduce the entire visible range of the spectrum. These different beams would produce independent holograms, which would be superimposed in the photographic emulsion. In theory, it should be possible to reconstruct a full-color image by simultaneously illuminating the hologram with all the colors used to create it, but in practice there's a long way to go. Russian holographers have made the greatest strides in this area.

RAINBOW HOLOGRAMS

Stephen A. Benton, a holographer at the Polaroid Corporation, has earned widespread recognition for his work on another type of white-light hologram. Unlike Denisyuk's, Benton's is a transmission hologram, and his trick was to *reduce* the amount of information stored in the hologram. When an ordinary hologram is reconstructed, it actually gives you more than enough information to see a three-dimensional image. Our sense of depth perception comes from looking at objects with both eyes, which are separated in a horizontal plane. But an ordinary hologram provides enough information to view an image in three-dimensional perspective from any angle. Benton devised an optical system that kept the horizontal perspective but lost the vertical. You can see this kind of holographic image as long as you hold it upright and don't tilt your head sideways. If you turn the hologram on its side, the image vanishes. More important, the design lets the hologram reconstruct an image using any wavelength of light in the visible region (see diagram 13).

Benton-type holograms are bright and are easy to reconstruct, since they only need white light. Their structure also causes an interesting side effect: a rainbow of colors, arrayed vertically on the reconstructed image. The colors change as you move your eyes up and down; they aren't realistic, but they can be striking.

HOLOGRAPHIC MOVIES

The next step is moving holograms, and it's a giant one. As usual, what you see in the movies isn't an accurate depiction of what you can expect. "Despite the expectations engendered by such popular films as *Star Wars,* these [moving] three-dimensional images will not be projected into thin air, but onto a 'screen' which is itself a complex optical element," Benton wrote in a 1980 paper on holographic imaging.

So far, the success in producing holographic "movies" has been far more modest than you might have expected from the scene in *Star*

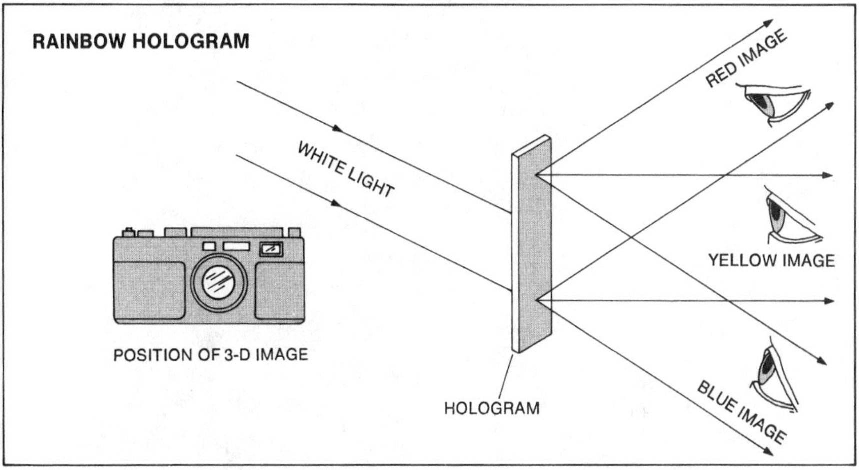

13. A rainbow hologram is illuminated by ordinary white light (such as light from a light bulb) rather than by a laser, although a laser is still needed to make the hologram to begin with. The rainbow hologram gets its name because it changes color as you look through the holographic film at different angles. This happens because the interference pattern recorded in the film diffracts different colors of light at different angles, breaking up the light much like a prism disperses white light into the rainbow of colors that make up the spectrum.

Wars in which the robot R2-D2 projected an image of Princess Leia (which, by the way, wasn't a real hologram, but a cinematic special effect). The most widespread approach, which uses a trick developed by San Francisco holographer Lloyd Cross, makes it possible to record only a limited amount of motion, and some holographers say that even then you have to stretch a point to consider the result motion.

The technique Cross developed, called multiplex holography, is an outgrowth of Benton's rainbow holograms. Just as a movie is basically a series of still photos, a multiplex hologram is formed by recording a series of parallel, vertical, narrow-strip rainbow holograms. Each strip hologram is the equivalent of a single frame of a conventional movie. The maximum capacity of a multiplex hologram is a little over a thousand strip holograms. The whole affair is recorded on a flat strip of film, which is then taped together at its ends to form a cylindrical hologram.

The best-known multiplex hologram is *The Kiss,* which shows holographer Pam Brazier blowing a kiss to the viewer. As you walk around the hologram, Ms. Brazier's eyes follow you. She also blows you a kiss and winks. Courtesy of the Multiplex Company

The cylinder is lit from inside with an ordinary light bulb. The image appears to be inside the cylinder and to be three-dimensional. As the cylinder is rotated, or as you walk around it, the image inside appears to move.

Multiplex holograms have been used most dramatically in "mini-movies," but they can also be used to make three-dimensional portraits, with images of a person shot from various angles around a full circle. The result is a single three-dimensional image inside the cylinder, which can be viewed from any angle.

Making a master multiplex hologram is tedious because of the many individual strip holograms that must be recorded. But it's easy to make copies of the master. The same is true of other rainbow holograms. This ease of reproduction, combined with the striking nature of multiplex holograms, has helped make them popular. Cross and a group of other holographers formed the Multiplex Company in San Francisco, and their work has circulated widely. The Multiplex Company produced the multiplex holograms of the head of actor Michael York that were used in a scene in the movie *Logan's Run* to portray York's interrogation by twenty-third-century police. The company's best-known multiplex hologram is *The Kiss,* which shows Pam Brazier, a holographer in her own right, blowing a kiss to the viewer. Copies in a variety of formats are available from many sources, including the Museum of Holography, the Edmund Scientific Corporation of Barrington, New Jersey, and the Holex Corporation of Philadelphia.

There are obvious limits. Because they are composed of rainbow holograms, images reconstructed from multiplex holograms show the same false colors. And because multiplex holograms contain only about a thousand "frames," they can reproduce less than a minute's worth of action at ordinary movie rates.

Practical true holographic movies appear to be a long way away. A Russian group is trying to reconstruct moving holographic images by projecting them on a special reflective screen, which in turn creates a separate image over each seat in the audience. The viewer must keep his eyes in a fixed position, and Benton, who's seen a demonstration, says it's "something like [looking through] a porthole to see the full-size three-dimensional image." In 1976, the Russian group showed a 20-second monochromatic film of a young woman arranging flowers, but only four people could watch it at a time.

Any kind of holographic movie would require coherent light, ruling out the possibility of making such films anywhere but in an indoor studio. Laser illumination of living actors raises serious questions of eye safety, so holographic movies would contain few searching facial close-ups.

WOULD YOU BELIEVE . . . HOLOGRAPHIC TV?

We wouldn't. Not for the immediate future, anyway. All the problems that plague holographic moviemaking—and more—also apply to any consideration of holographic television. To start with, you have to have a medium capable of transmitting all the information contained in holograms to homes. With recent advances in signal compression, it looks as if it may be possible to transmit holographic signals over the equivalent of about 500 color-television channels. That's at the upper end of the capabilities fiber-optic technology is expected to have in the near future (see chapter 6).

Then you'd need the holographic equivalent of a television camera, which would have to have a resolution of a couple of thousand lines per millimeter (some 50,000 lines per inch) in order to record a hologram properly. That's far beyond the capability of existing television cameras.

You'd also need some medium that could display the hologram in your home. It would have to change its transparency, or reflectiveness, in response to the signal being transmitted and would require the same resolution as the holographic camera—a couple of thousand lines per millimeter or roughly 50,000 lines per inch. Today an ordinary U.S. television has but 525 lines displayed from top to bottom of the screen (a European television has 625 lines).

Finally, you'd have to have a way to reconstruct the image from the holographic screen. Because reconstruction is very sensitive to the viewing angle, you'd only be able to sit in certain places in your living room. But that would be one of the smaller problems.

Holographers hesitate to call anything impossible, and there's still some interest in holographic television. One holographer says there's a chance our grandchildren might live to see it. Another reportedly said he could solve the problem in a year or two—given the Air Force's entire budget during that time.

ART, GAMES, ADS, PORNOGRAPHY

Holography is used most visibly as an art form and for making various displays. The three-dimensional nature of holographic displays grabs attention, important for such things as advertising. Holograms make interesting novelties, whether as pendants or hung on the walls of corporate lobbies. A few small companies are busily producing holographic jewelry, and Stephen Benton estimates that "almost a million Americans have holograms in their homes or offices, or around their necks. . . ." Millions more have seen a hologram, either directly, as at the Museum of Holography or one of its traveling exhibitions, or indirectly, as in *Logan's Run.* Others *think* they have seen holograms, but what they saw was really a three-dimensional image produced another way, as at Disneyland's Haunted Mansions or in *Star Wars.*

Holograms have also entered the world of electronic games. A tabletop game called *Cosmos* has been developed by Atari of Sunnyvale, California. Interchangeable cartridges are used to program the game's several variations, and each of these cartridges contains two holograms. When the game is being played, light-emitting diodes (LEDs) reconstruct a holographic image of background scenery on a 3½-by-4-inch (90-by-100-mm) screen. One image is displayed through most of the game, but near the end the display automatically shifts to the second hologram. For example, the *Space Invaders* version shifts from a three-dimensional moonscape to an extraterrestrial monster at its climax. The game was originally scheduled to go on sale in the last half of 1981 for under $100, and individual cartridges were to be priced at about $10—an impressive achievement, if true, for modules that contain two holograms as well as electronic elements. However, the company has delayed that introduction for internal corporate reasons.

Cosmos isn't the first game to use holograms. That distinction apparently belongs to *Gunsmoke,* an arcade game (the kind you put money in to play, as opposed to a home game) that includes a multiplex hologram of a gunfighter. The game opens with the gunfighter challenging you to a duel. If you outshoot the gunfighter, the hologram is turned one way, and you see your opponent fall to the ground. If you are too slow on the draw, the hologram will turn in the opposite direction, and the gunfighter will draw his gun, shoot, and sneeringly reholster it. The machine then will announce, "You lose!"

About 3,000 games of *Gunsmoke* were made for Kansaiseiki, Ltd., a Japanese gamemaker, and another 750 for Midway, a Chicago-based U.S. game company. While the makers of arcade games are large organizations, development and production of the holograms remains something of a cottage industry. The game was designed by Conceptioneering, a tiny Santa Clara, California company, whose president, Leonard Gesensway, portrays the gunfighter in the hologram. The holograms themselves were produced by the Multiplex Company.

Holography has also been used in more blatantly commercial displays. A multiplex hologram was the centerpiece of a display case advertising Salem cigarettes in New York City's Grand Central Station. And there are several pornographic multiplex holograms in circulation. They're generally of the soft-core variety (we've seen a hologram of two nude women fondling each other), and appear at such places as engineering-oriented trade shows, where they're used to draw the attention of the largely male audiences—not merely because of their subject matter but also because of their technical novelty.

While advertising and pornography may pay the bills, the first love of many holographers is art. The holographic artist unfortunately has a hard row to hoe. Few artists have any technological training, yet making good-quality holograms requires sophisticated technological know-how. It also requires expensive hardware, and most holographic artists suffer from the time-honored poverty endemic to the art world.

To cope with these problems, holographers have developed some relatively inexpensive ways of producing holograms. Instead of working on sophisticated optical tables, for example, many work in sandboxes. These aren't the ordinary kind in backyards, which are full of toys and small children. Their design, credited to holographer Gerry Pethick, involves placing washed sand in brick boxes supported on motorcycle inner tubes. The optical components are held in plastic drainpipes stuck into the sand. The inner tubes isolate the table's surface from floor vibrations, and the sand provides weight and additional stability without transmitting vibrations on the surface.

The surroundings are often humble. Dan Schweitzer and Sam Moree operate the New York Holographic Laboratories in the basement of a movie theater at 34 West 13th Street in Manhattan. The noise from the theater can sometimes cause problems, but the artists cope by taking such measures as turning off the air conditioning while exposing holograms on their sandbox equipment. The Museum of Holography has its own basement laboratory with more sophisticated equipment.

Some of the results are striking not merely for their novelty but also for the artistic vision that underlies them. In his *Twelve Milliwatt Boogie,* for example, New York holographer Rudie Berkhout superimposed three rainbow holograms to produce a multicolor image of several geometric figures. The float in space before the eye, shifting colors as the viewer moves his head up and down, and at times becoming distorted by the limitations of the holographic recording. Berkhout calls the distortion-induced effects "hyperspace," and tries to use them creatively.

A walk through the Museum of Holography makes you realize the diversity of holographic visions. Among the artists who have had one-person shows are Harriet Casdin-Silver of the Center for Advanced Visual Studies at The Massachusetts Institute of Technology, Ruben Nuñez of Venezuela, Carl Frederik Reuterswald of Sweden, and Anait Stephens of California.

One of the Museum staff's biggest problems is mounting the holograms to get the best effects, as the quality of the reconstructed images is very sensitive to the type of illumination used. For instance, one of us has a rainbow hologram of a carnation produced by San Francisco artists Randy James and Gustavo Houghton that is impressive when illuminated at just the right angle by full sunlight. Without this lighting, the image seems shallow and fuzzy.

Besides being a creative medium in itself, holography is also being used to preserve other great art by detecting stresses and strains in old paintings or their substrates. The technique is called *holographic interferometry.* Besides aiding art historians, it's the most important industrial use of holography, and we're going to have to take a detour to explain just how it works.

HOLOGRAPHIC INTERFEROMETRY: TESTS IN THE THIRD DIMENSION

We mentioned earlier how electromagnetic waves can interfere with one another to produce a pattern of light and dark fringes. In chapter 9 we described how interference can be used to measure short distances—even smaller than the wavelength of light. Ordinary interferometry measures distances between objects, but holographic interferometry measures distances between points on reconstructed holographic images.

Let's say we're inspecting an airplane tire. First we take a hologram of the tire under normal conditions. Then, leaving both the tire and the holographic plate in the same places, we blow hot air against the tire. Then we take another hologram and record it on the same plate.

The result is a double-exposure hologram. It's really a superposition of two holograms recorded on the same photographic plate. When reconstructed, both images are produced, superimposed on each other. Because the tire was kept in the same place, you see a single image of the tire. However, superimposed on top of the tire you see an interference pattern of light and dark lines, which shows how much the size of the tire changed as a result of the heat. If there's a minute flaw in the tire, it shows up as a spot where there are many interference fringes. The technique is very successful for testing aircraft tires. In fact, it's the only method so far that's met tire-testing requirements laid down by the Federal Aviation Administration.

The same principle can be applied in countless other types of nondestructive testing—so called because the tests are designed not to harm the objects being tested. For example, the strength of a piece of welded metal that is bent slightly can be tested with double-exposure holography, by comparing the metal before and during bending. Holographic interferometry has been used to measure fluid and gas flow, the effects of orthodontic braces (as mentioned in chapter 5), and even the vibrations of old Chinese musical instruments.

Where does art come in? Because holographic interferometry is nondestructive, it can test how well delicate old works of art are surviving. For example, a team from the Universities of Aquila (Italy) and Rome used holographic interferometry to observe separations between layers of paint in a fifteenth-century wood-panel painting. They took double exposures: the first of the painting in its normal condition, the second when warm air was blown across its surface to heat it. The result was a set of regularly spaced fringes where the paint was adhering properly to the wood, with deviations indicating where the paint layers had separated.

INDUSTRIAL HOLOGRAPHY

Holographic interferometry appears to be the single biggest use of holography in industry. Yet industry doesn't like to talk about it. To understand why, we need to take a brief excursion into history.

The development of laser-based reflection holography in 1963 set the stage for the discovery of holographic interferometry, and in 1963 and 1964 about half a dozen research groups independently and almost simultaneously discovered the phenomenon.

One result of the flurry of holographic research in the early 1960s was that a number of patents were issued to researchers, including Leith and Upatnieks. Many of these patents were later acquired and assembled into a package by Holotron, Inc., which was set up in Columbus, Ohio, by the Battelle Memorial Institute and the Du Pont Company. For several years Holotron tried to get holographers to pay for using the patented technology, but to little avail. Eventually, Du Pont and Battelle sold Holotron to a small company, Holosonics, Inc., of Richland, Washington, which also had little luck in enforcing the patents. Holosonics had little luck in anything else and eventually went bankrupt, leaving the patents, as the prime assets, in the hands of the bankruptcy trustees.

There have been some questions raised about how strong the patents are—they've never been tested in court, apparently. But the basic problem Holotron and Holosonics had was tracking down people using holography. The people selling holograms are generally artists or small companies, and few of them sell enough holograms for a lawsuit to be worth the trouble.

Industry wasn't buying holograms, but it was buying optical hardware to make holograms. Buying the hardware didn't say much, however, since the same equipment could be used for purposes other than holography. Moreover, holography could be done quietly, using company facilities, with no one at Holotron or Holosonics any the wiser. With the possibility of a lawsuit charging patent infringement looming in the background, few companies that use holography talk much about their techniques or what they use them for.

At this writing, the Holosonics trustees have begun selling "exclusive licenses" for using the holographic techniques described in the patents for various applications. The license for laboratory and nondestructive testing uses (which includes holographic interferometry) was sold to the Newport Corporation of Fountain Valley, California. Newport, which makes holographic equipment, hopes it will be able to make some money by sublicensing the patents at modest rates to industrial holographers. Newport will also be free to offer complete systems for industrial holography without having to worry about patent disputes. The company's motives aren't purely business-related, though. Talking with Newport vice-president Milton Chang, you sense that he's glad to be helping holography along. He, as well as the company's president, principal scientist, and advanced projects manager, did his doctoral research in holography at the California Institute of Technology.

The increasing pressure of quality control may help holography find more of a home in industry. So will new technological advances, such as a new type of holographic film that develops itself within seconds and can later be erased and reused. Such a film is ideal for routine inspection by holographic interferometry, as long as it's not essential to keep an archival record of the tests.

ARTIFICIAL HOLOGRAMS

In theory, holography is a tremendously powerful optical technique, because it lets you generate whatever type of wavefront you want. So far we've talked mostly about simply preserving a wavefront, so that it can later be reconstructed to produce an image. But that's just the beginning of holography's capabilities.

Scientists sometimes look at holography as being the equivalent of a complex mathematical function called a Fourier transform (see chapter 11). The details are unimportant here; what is important is the idea that it's possible to describe mathematically a hologram and the object it represents. This mathematical model enables a computer to synthesize holograms. Such holograms may be designed to represent a specific object or, more often, to represent a particular mathematical operation on the light passing through them.

Holograms can also be made to perform the functions of complex lenses, mirrors, and other optical components. A practical example is a holographic scanner, developed by the IBM Corporation, which can be moved in such a way that it directs a beam of light in a special pattern to read information coded in a series of bars (called the Universal Product Code, or UPC) on packages in supermarkets (see chapter 11). A holographic lens can be used to project information into the line of sight of pilots flying military aircraft, so that they don't have to take their eyes off what's going on outside in order to read their instruments. The holographic lens serves the function of a regular lens, but one that wouldn't otherwise be practical to make.

Neither computer-generated holograms nor holographic optical elements have found wide applications at this writing. Both have practical flaws but also offer exciting prospects if these flaws can be overcome.

NONLASER HOLOGRAPHY

Obviously, we have concentrated on holography using lasers, which are the best light sources for the process. However, holography is a general principle that can work with all types of waves, not just light.

For instance, holograms have been made using microwaves. Holography can also be performed with ultrasonic waves (ultrasound). Such acoustical holography has a variety of applications in research and nondestructive testing.

Remember, Dennis Gabor's original goal was to improve the resolution of the electron microscope! If you've been reading carefully, you may wonder what Gabor thought he was doing when he applied a wave process to a beam of electrons. It turns out that electrons do have some wave properties, just as electromagnetic waves have some of the properties of particles. It's possible to diffract electrons, and diffraction is the basis of the reconstruction of a hologram. Thus it's possible to freeze an electron wavefront in time. Ironically, the original application that Gabor envisioned has yet to become practical.

13 THE LASER AS ARTIST AND ENTERTAINER

The uses of lasers in art and entertainment aren't limited to holography. Laser art can also be bold, brightly colored beams in vivid contrast to a dark background. Turn the artist's paints into beams of light, and let his imagination run riot; the laser is coming.

Actually, the laser may already have come—at least if you live near a large city. Laser light shows have played in planetariums and theaters the world around; there was one at President Reagan's inauguration. By the end of 1980, over nine million people in places from Caracas to Calgary had seen a single series of laser light shows—the *Laserium* shows produced by Laser Images, Inc., of Van Nuys, California, the biggest, and probably the best, show-producer.

Lasers can also play a quiet role in entertainment, for lasers are one of three competing technologies that can be used to play videodisks, which may be the next stage in the video entertainment revolution. Magnavox and the Pioneer Electronics Corporation are selling videodisk players that contain helium-neon lasers, which are used to play back television signals recorded on reflective disks. The process is similar to using a phonograph needle to play back music recorded in the form of grooves on a record, but it's hundreds of times more demanding.

Before you get carried away, remember that the world of entertainment is a world of hype. There are cynics who can make a convincing case that the videodisk is high technology's answer to the Edsel. Some light shows have been dismal failures; others have endangered public safety by shining beams into the audience. And there are artists who wonder if laser art is merely a gimmick that attracts attention more because of the laser than because of its artistic merit.

As is inevitable, reality lies somewhere between the two extremes. Laser beams can be striking tools in the hands of a talented artist—yet technological constraints limit what can be realized in laser art. Light shows can be spectacular—and they can also pale with age and repeated viewings. And the returns that count—from the marketplace—aren't in yet on videodisks.

LASER AND KINETIC ART

Most laser art is basically a subcategory of kinetic art, in which motion is integral to the artwork. The motion can be produced in many ways; often it is in response to music. By its very nature, kinetic art tends to be highly technological, and it was only natural that kinetic artists turned to lasers.

It wasn't until the late 1960s that many artists began serious efforts to incorporate lasers into kinetic art. The problem was that of the early types of lasers, only the now ubiquitous helium-neon laser produced a continuous beam of a type that could be useful. While the helium-neon laser was (and still is) relatively inexpensive, the dull red beam it produced wasn't very exciting artistically. Then, in the late 1960s lasers containing the rare gas krypton came onto the market. Krypton emits several bright lines in the visible region and for the first time gave the laser artist a palette of colors.

We mentioned earlier the economic problems faced by holographers, but the difficulties are even more severe for laser artists who want to work in color. You can set up a working holographic system for a few hundred dollars, though spending more money will often make it possible to produce better holograms. Prices for krypton lasers (and for mixed-gas lasers that contain a mixture of krypton and argon) start around $10,000, and that doesn't include any of the equipment to deflect and modulate the beam to make interesting patterns. Because the hardware is so expensive, multicolor laser art isn't within the reach of the poverty-stricken artist; it generally requires external financial support or has to be put into the format of commercial entertainment with an admission fee.

THE PEPSI PAVILION

One of the earliest showings of a full-color laser artwork was at the Pepsi-Cola Pavilion at Expo '70 in Osaka, Japan. The laser display, which was literally in the basement of the Pavilion, was only a part of a fascinating experiment in technological art called "Experiments in Art and Technology," conducted by a group of seventy-five artists and engineers. The story of the Pepsi Pavilion is a book in itself [*Pavilion,* by "Experiments in Art and Technology," Dutton, New York, 1972], but here we will focus only on the laser display.

The idea of a laser show was conceived by the composer David Tudor, but the actual system was designed by Lowell Cross (now at the School of Music at the University of Iowa in Iowa City) and Carson Jeffries, a physicist from the University of California at Berkeley. Their idea was first to split the beam into its four component colors, then to use music to control the electronics that caused pivoting mirrors to turn and steer the laser beams around the room. There was a separate set of mirrors for each color.

Getting the laser system up and running proved to be an adventure. When the laser from Coherent Radiation Laboratories (now Coherent, Inc.) of Palo Alto, California, arrived initially, the light-producing tube of krypton gas inside the laser was broken. A second tube shipped from the factory to Japan also arrived broken. A third tube arrived intact the day before the Pavilion was to open, and Cross and Elsa Garmire—a laser physicist then at the California Institute of Technology, but now at the University of Southern California—worked until after midnight setting it up. They were hard at work again the following morning, and they finally got the laser working perfectly at 12:50 P.M., 10 minutes before the dedication of the entire Pavilion (of which the laser display was only a part) was to begin.

The result was an interlacing pattern of red, yellow, green, and blue lines, which flowed across about half the room where the beams were projected. The light played over the faces and clothes of visitors, apparently without ill effect. The result was striking. One art critic called it "an electric [Jackson] Pollock," after the artist best known for strewing blobs of paint about to create patterns.

LIMITATIONS

A decade later, Cross and Jeffries are still working together, still using music to control the motion of laser beams while seeking to realize the same type of artistic visions. Artistic vision seems to be the key need now, Cross thinks. He'd like to do more with computer control systems, but there have been few other major advances in the relevant technology in recent years.

The krypton laser itself remains expensive and delicate, although there have been some improvements over the years. Cross tells horror stories of shipping the fragile laser system for shows away from his Iowa City base. Once, he found the crate containing the laser standing on its end, despite vivid warnings on the crate to keep it horizontal. That mistake cost the freight-handling company the price of a new tube for the laser.

Cooling has long been a critical concern in any display that uses krypton or mixed-gas lasers. One laser artist reportedly left a trail of dead artwork behind him, because he wasn't careful enough in meeting cooling requirements. The higher-power krypton lasers generally require flowing water for cooling; at lower powers it's often sufficient to blow air through the laser with a fan. Cooling is not critical for helium-neon lasers, which are used in some laser artwork because of their low cost and simple reliable operation.

The key technological limitation on laser art, however, is not the laser itself but the way in which the beam is deflected. There are several ways to deflect laser beams, but the best choice for light shows is generally a pivoting mirror. This mirror is mounted on a rod that turns back and forth on its axis, sweeping out an angle of up to about 30 degrees. A laser beam reflected from the mirror scans the same angle. To trace out a two-dimensional pattern, the beam is deflected successively by two scanning mirrors. Usually the light from a krypton laser is separated into its component colors before it goes to the scanning mirrors, and a separate pair of mirrors is used for each color.

The pivoting of the mirrors is controlled electronically, but the motion of the mirrors is mechanical, and like all such motion it's subject to mechanical limitation. Even the best mirrors can scan back and forth no more than about 2,000 times per second. That's fast enough to fool the eye into seeing the pattern, rather than the moving laser spot. But it's not fast enough to keep a complex pattern in the viewer's eye without part of it fading. The laser beam thus has only enough time to draw a very simple pattern before it has to go back and redraw the pattern so that it doesn't fade from view.

You can see the limitations of laser scanning when you stop to compare it with a television tube. The beam of electrons that writes a signal on a black-and-white television tube scans the screen more than 15,000 times per second (30 separate frames of 525 lines each), compared to the laser's 2,000 times per second. Moreover, the television tube contains phosphors that continue to emit light even after the electron beam has gone its way, producing what looks to the eye like a continuously illuminated picture, although in actuality a bright spot is simply scanning the screen over and over. The speed limitation is why lasers can at best trace out simple line drawings but never the photographlike images you see on television.

Even line drawings can be a strain on the scanning mirrors. The maximum scanning frequency is audible, and at some laser displays you can hear the mirrors humming, straining to scan back and forth as fast as they can. The quality of cartoonlike images is also limited by another problem, which engineers call "ringing." This means that the mirror doesn't stop precisely where it's supposed to, but instead vibrates slightly, causing the laser spot to waver on the screen.

Although these constraints seriously limit what laser artists can achieve, artists working within them can still create some interesting visions. If the scanners are driven at a fairly constant frequency, rather than being controlled to draw specific patterns, they can readily trace out fascinating and apparently complex images. Some of the images are similar to those that can be traced out using a child's Spirograph set of notched wheels and rings, and the phenomenon involved is somewhat related. Simultaneously driving the two mirrors at different frequencies causes the laser beam to trace out a repetitive pattern that depends on the relationship of the two frequencies—just as the pattern traced out by the Spirograph wheels depends on the number of teeth on each notched edge. Such patterns are the hallmarks of many laser artists and of such light shows as *Laserium;* they are often produced by, or synchronized with, music. They're also relatively easy to produce.

PROFILES IN LASER ART

Although artists are individuals who approach laser art from different directions, some commonalities have evolved. In some cases the convergence seems to stem from the technological limitations, in others from a common artistic vision.

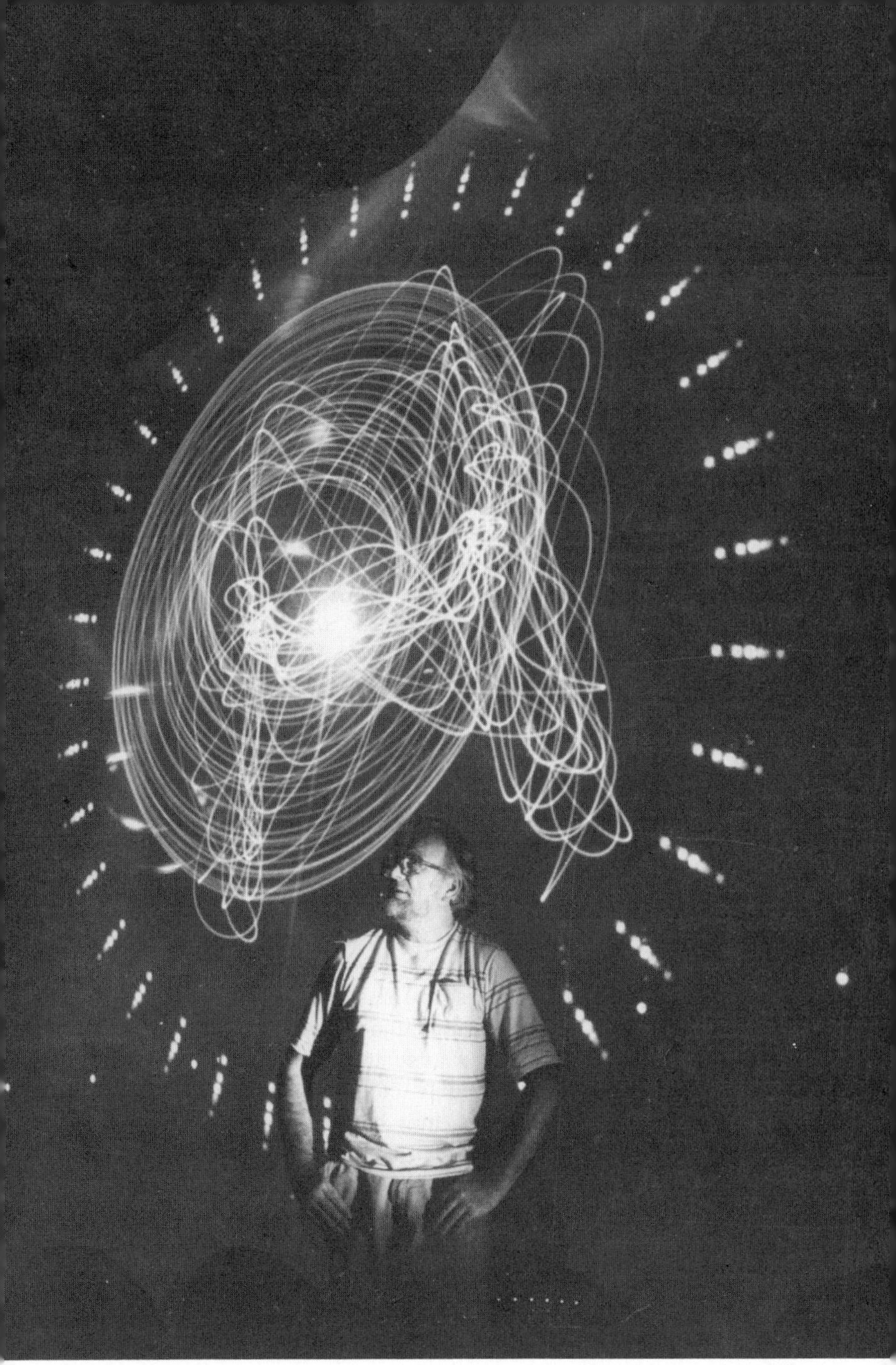

J. Stanislaus Ostoja-Kotkowski is seen during the Adelaide (Australia) Festival of Arts in 1980 with one of his music-controlled laser creations. Courtesy R. Nicholson

The starting point for Cross, a musicologist by training, was electronic music. He became interested in the interaction of light and music and experimented with the use of various types of music to drive laser scanning systems. He feels that symphony orchestras provide the best music for driving laser art and reports his best audience successes with orchestral music. Although most light shows use a large proportion of rock music in their sound tracks, Cross feels that rock "won't lead too far artistically." He's also looking at computer techniques for controlling the lasers.

Another artist interested in the interaction of light and music is J. Stanislaus Ostoja-Kotkowski, a Polish-born artist who lives in Stirling, South Australia. He describes himself as a painter who uses the laser as a brush. Most of his works are kinetic pieces, which combine scanning laser beams (often driven by music) with "washes" of color from incandescent lamps. His laser creations have appeared at many art festivals in Australia and elsewhere. One of his more striking works was the *Laser Chromasonic Tower* built for the Australia '75 Arts and Sciences Festival in Canberra. Incandescent lights and lasers driven by music scanned across the translucent walls of a tower, which was placed in a shallow water basin to add reflective effects. In the evening the bright lights on the tower walls stood out boldly against the surrounding blackness.

Although their artistic visions differ, Cross, Ostoja-Kotkowski, and many other laser artists have much in common. Those working in large areas typically use argon, krypton, or mixed-gas argon-krypton lasers. Those working in smaller spaces (or sometimes just with smaller budgets) may be limited to red-beam helium-neon lasers. They invariably use some sort of mechanical arrangement to scan the multicolored beams. Various means are used to modulate the intensity of each beam. Devices called "choppers" may block the entire beam some of the time, but in other cases the intensity can be controlled over a continuous range. The beams are projected against clouds, screens, special structures, or even intentionally produced fogs.

In general, laser artists are interested in the kinetics of the laser beams. Typically the motion of the beams is under the direct control of the artist, who may either be sitting at a console or have generated a program tape to control the motion of the mechanical mirrors. Often this control is synchronized with—or even operated directly by—music.

LASER ENTERTAINMENT

Some of the ideas that have intrigued laser artists have been commercialized into a subgenre of the entertainment industry—laser light shows. The biggest and best of the commercial laser light shows is *Laserium,* produced by Laser Images, Inc., of Van Nuys, California. There have been several generations of *Laserium* shows, and paid attendance through 1980 had surpassed nine million people.

The origin of Laser Images, Inc., goes back to 1970, when Elsa Garmire, having returned to the California Institute of Technology from her work on the Pepsi Pavilion in Japan, introduced a young filmmaker named Ivan Dryer to laser art. Dryer's first idea was to try to use laser-generated images in a film. When that didn't work, he decided to try a live laser show.

Dryer founded Laser Images in early 1971, and for a couple of years the company struggled along quietly. The company's first hit was the original *Laserium,* which opened at Los Angeles's Griffith Planetarium. Since then, the company has grown to include 110 employees, with shows presented on a continuing basis in twelve cities around the world (Denver, Las Vegas, London, Los Angeles, New York, Philadelphia, St. Louis, San Francisco, Seattle, Tokyo, Toronto, and Washington). The show has gone through several versions and has toured or had temporary engagements in many other cities around the world. The shows are often presented in planetariums, because the spherical domes make perfect backdrops for the laser images.

The heart of the *Laserium* show is a projector, which includes a krypton laser, driving electronics, scanning mirrors, controls, and a four-channel tape deck. Two channels on the tape deck play stereo music; a third carries the voice of the narrator. Information needed to control the projector and synchronize it with the music is recorded on the fourth channel.

One of the reasons why *Laserium* has been so successful is that each projector is operated by a "laserist," who's intimately familiar with the scanning laser system. The laserist sits at an electronic console that allows him to modify the patterns created by the projector. Most laserists are musicians, and their function is somewhat akin to that of sound engineers in recording studios or at concerts. Use of a laserist lets Laser Images, Inc., avoid the problem of repetition inherent in a completely preprogrammed show. Each *Laserium* concert is unique.

Laserium concerts have sometimes used classical music—the original *Laserium* show included Johann Strauss's "The Blue Danube." But rock music has been dominant, with songs ranging from such oldies as Chuck Berry's "Roll Over Beethoven" and Elvis Presley's "Heartbreak Hotel" to contemporary songs such as Dire Straits' "Sultans of Swing" and The Cars' "Just What I Needed."

Most other laser light shows follow a similar pattern, with a heavy emphasis on rock music. A major reason is commercial; rock music tends to draw heavier crowds than other types. Rock concerts were also the first to be accompanied by light shows, back in the psychedelic 1960s, and some of the songs from that era show up on *Laserium* play lists, notably "Light My Fire" by the Doors, "White Rabbit" by the Jefferson Airplane, and "Purple Haze" by Jimi Hendrix. The continuity with the psychedelic era is visible in other ways as well, such as in the occasional sweet smell of marijuana smoke at a laser concert. But of late, some laser shows have encountered a different problem—disorderly beer-drinking youths, who caused enough problems for Boston's Hayden Planetarium to discontinue laser shows.

Although the *Laserium* format, in which rock music has largely provided the backdrop to the laser show, has been the most successful, there have been other successful types of shows. *Lovelight,* a joint production of the Intermedia Systems Corporation and General Scanning, Inc. (the major supplier of scanning mirrors), was written as a musical play and included some cartoonlike images as well as more abstract forms. The star projector at Boston's Hayden Planetarium provided a stellar backdrop for some of the show's laser images. The plot was sketchy—reminiscent of the image sequences of the latter part of Stanley Kubrick's *2001*—but audiences found the concept interesting, and the show played for six months in Boston.

Laser shows or displays have a tendency to pop up all over. There were laser shows throughout the summer of 1980 as part of Boston's 350th birthday celebration, and, as we have already mentioned, there was one in Washington, D.C., to celebrate President Reagan's inauguration. A laser light show was even part of the United States' bicentennial celebration. On July 4, 1976, Soleil—a light-show company organized by Bruce Rogers and Gary Levenberg—put on a laser show from the top of the Washington Monument. The show, an official part of the bicentennial, was seen by an audience estimated to have numbered four million people, and the beams were reportedly visible as far as 20 miles (about 30 km) away.

Laser shows have also been put on in countless other locations. During the heyday of the disco fad, a number of discothèques set up their own laser displays. The trend was widespread enough in France that during 1979 some seventy argon, krypton, and mixed-gas lasers were installed in French discos—more than were purchased by French scientists during the same period.

The demise of the discos may be a disappointment to some laser light-show producers, but there's still lots of interest in laser displays. Laser Images, Inc., for example, has a long list of credits, including motion pictures, television programs, live concerts, debuts of record albums, trade shows, outdoor laser shows, and presentations at amusement parks, such as Knotts Berry Farm and Magic Mountain. There are many other companies that produce laser displays; their names range from Laser Displays, Inc., of Boston, to the Science Faction Corporation, of New York. In Switzerland, Skyliner Promotion AG of Binningen offers to rent a van equipped with a complete assortment of light-show equipment for about $7,000 a day and to sell advertising time in laser shows for about $10 a second.

ROCK GROUPS AND SAFETY PROBLEMS

Among entertainers, some of the most enthusiastic response to laser shows has come from rock-music groups. The British band The Who has probably made the biggest commitment, supporting development of light-show and holographic technology at a company called Holoco, Ltd. which did special laser effects for the movie *Outland.* Other groups have generally been interested more in putting on a spectacular show than in the nuts and bolts, and that's caused some problems, because many of the musicians haven't understood that too much light, like too much sound, can be harmful.

There are a handful of horror stories that have made the rounds of the laser community. One concerns a group of young British laser specialists, who carefully built a laser system and equipped it with safeguards for a French rock concert. Then they watched in horror as the rock group's roadies removed the safeguards, allowing the beams to scan the audience during the show. After the concert, the Britons hurriedly loaded their equipment into a van and sped off for the Belgian border, convinced that the French police would not be far behind them. They were evidently unaware of the lack of French regulations at the time; it wasn't until the spring of 1979 that France established its first laser safety rules. Now French police are supposed to check each laser show before it begins.

Beams have also scanned through audiences at a few U.S. rock concerts. A member of one rock group wore a mirror on his wrist and used it to scan a multiwatt beam across the audience. There have been reports of other groups doing similar things.

Remarkably, there have been no reports of injuries inflicted by such carelessness. There's really no possibility of skin burns from such lasers, but it is possible for such intense beams to cause permanent eye damage. A 1-watt laser beam is over a hundred times as intense as sunlight, and it's possible for it to inflict a permanent blind spot on the retina. The fact that such injuries have not been reported at light shows may be due to pure luck, overly conservative standards, or simply the difficulty in noticing a small blind spot.

In an effort to curb such dangers, the Bureau of Radiological Health (BRH), an arm of the Food and Drug Administration, has come down hard on operators of laser light shows and has established stringent safety regulations. The BRH regulations appear to have controlled the most hazardous practices, which, in general, were due to ignorance rather than malice. However, there have been some unfortunate side effects of the regulations, most notably a plethora of paperwork for legitimate producers of laser light shows. The paperwork requirements have made it almost impossible to put on a series of one-night shows in different locations, and some makers of light-show equipment are complaining that BRH is being unrealistically strict. Other members of the laser community have made similar criticisms of BRH and have said that the agency's rules are based on inadequate data; often such people also complain that it was carelessness on the part of a few rock groups and light-show operators that brought about what they consider to be bureaucratic overreaction. Others say it's simply bureaucratic empire-building.

WHAT'S NEXT?

At the moment, many laser artists and producers of light shows consider overregulation by BRH to be their biggest single problem. However, there's another important problem that has nothing to do with safety or regulation: the technology for laser art and entertainment has reached a plateau.

"There are no technological breakthroughs on the horizon," says Lowell Cross; innovations will have to come from the visions of the artists. Laser artists are drawing on a very limited range of technologies. One result, in Cross's words, is that all laser shows "tend to look the same. Once you get over the initial dazzle, many effects look similar to one another." Indeed, he feels that the audience's real interest in laser shows is often in the laser itself and what it's doing. That attitude is common among those who put on light shows, who often go to great pains to insert the word *laser* in the title. *Lovelight* was almost unique in not using *laser* in its title, but when a new edition of the show was designed, it was given the title *Laser Magic,* much to the disgust of some of the staff.

Cross himself is looking in other directions. He's never been satisfied with using rock music to drive laser shows, perhaps because he doesn't like rock. He's experimenting with computer control of the laser system and is trying to create what he calls "pseudo-three-dimensional" nonholographic images, which would fool the eye, although they would not actually be three-dimensional. He is also experimenting with writing letters and numbers and drawing cartoonlike images, and is paying particular attention to the difficult problem of drawing cleaner figures without extraneous lines or distortions.

TOOLS FOR ARTISTS AND ART RESTORERS

We talked about how lasers could be used in various types of materials working in chapter 8. Some artists also do materials working, although they call it by such names as sculpting, engraving, or etching. Lasers can do some of these things and in a few cases have done them. But lasers capable of such tasks are expensive tools, and few artists can afford them, even if they are willing to give up the "feel" of creating artwork with their hands. However, lasers do etch wood and plastic for mass production of gift items, because they can repeatedly cut complex patterns easily and rapidly.

Probably the most fascinating artistic use of lasers for materials working is in the restoration of deteriorated artwork. Much of the work involved in art restoration lies in removing material, ranging from dirt and corrosion to paint and other materials added in previous restoration attempts. Scraping off this material mechanically or etching it away chemically is a painstaking task and runs the risk of seriously damaging the already fragile artwork.

As an alternative, John F. Asmus, a physicist at the University of California at San Diego, has developed a laser technique. He has used pulses from ruby or neodymium lasers to remove black encrustations that were accelerating the decay of marble sculptures in Venice that date back to the Renaissance. He has also used lasers to remove hardened and moldy glue from the back of a painting, fungus from a leather-bound book, and tarnish from silver leaf in textiles. Lately, Asmus has used lasers to restore paintings made 2,500 years ago by North American Indians.

THE LURE OF THE VIDEODISK

Lasers are playing a bit part in one of the most fascinating adventures in home entertainment technology and marketing in recent years—the videodisk player. If you believe the promoters, videodisks will start the home entertainment revolution of the 1980s. If you believe the cynics, the videodisk will be high technology's answer to the Edsel, doomed to the same oblivion as quadraphonic sound. There's something to say on both sides of the issue.

The basic idea of the videodisk is simple: a videodisk is the video equivalent of a phonograph record. Instead of playing back music as a phonographic record player does, a videodisk player plays back a video program on a television set. The goal is a video equivalent of the phonograph, with video records (videodisks) that are simple to mass-produce and players that are simple to use.

The idea is simple, but the technology is difficult. A phonograph needle follows a single spiral groove around and around a record, and the mechanical motion of the needle reproduces the recorded music. This works fine for music, where it's only necessary to reproduce frequencies up to about 20,000 hertz. To reproduce a color television signal, on the other hand, you have to reproduce frequencies up to about 6.3 *million* hertz. You can't do that with a phonograph needle.

Engineers went to work on the problem and came up with a variety of approaches. Most of the approaches didn't work, among them a scheme for recording holograms on a special tape and playing them back. One that did, after many years of effort, involved replacing the phonograph needle with a beam from a helium-neon laser and the record groove with a spiral pattern of reflective spots. It was an elegant solution to the problem. There was no need for mechanical contact between the disk and the "needle," and hence there was no wear. The video information recorded on the disk was embedded where, at least in theory, it couldn't be harmed by scratches or fingerprints. The laser beam could be focused to a tiny point, making it possible to encode the video signal in a series of points so small that half an hour to an hour of video program could be stored on a single disk.

The laser approach was developed independently by N.V. Philips, the Netherlands-based electronics giant, and MCA, Inc., a U.S.-based producer of motion pictures. Eventually the two companies merged their programs and specialized in their efforts: Philips now produces players at its Magnavox subsidiary in the United States; MCA produces disks and has secured rights to distribute motion pictures and other material on videodisks. Several other companies have allied with Philips and MCA, among them IBM (which is a partner in MCA's disk-making venture, DiscoVision Associates) and the Japanese Pioneer Electronics Corporation (which markets its own videodisk player).

The lure of the laser is not lost on videodisk marketers. Pioneer calls its system the "Laserdisc." Some stores use optical videodisks in displays, because the closely spaced spots on their surfaces create a rainbow diffraction pattern.

It turns out, however, that you don't have to have a laser to build a videodisk player. The RCA Corporation has developed a videodisk player in which the "needle" is an electronic sensor of patterns in the disk. A similar scheme has been developed by the Japanese Matsushita Corporation, but you can't play disks for one player on the other. At this writing, the RCA system has just gone on the market, and Matsushita is gearing up for an introduction sometime in 1982. The battle lines are forming, and each company has its own set of commercial allies involved in making compatible equipment.

It now looks as if the major competition may be between the RCA and MCA/Philips groups. Each side has its own advantages. At about $500, the RCA player is about $200 to $300 cheaper than any laser-based videodisk player, with disk prices also slightly lower. RCA also has a strong dealer network, including Sears Roebuck and Company. However, the laser-based players offer broader capabilities, and they've been on the market longer. Both sides have put lots of money into videodisks—RCA is estimated to have spent over $150 million; MCA and Philips probably haven't reached that total, but their spending can't be far behind.

At the end of 1980, both Philips and RCA were issuing optimistic statements, which is only natural, given how much money they have at stake. The chairman of Philips said that he expected "hundreds of thousands" of laser videodisk players to be sold in the United States in 1981. RCA predicted that it would sell 200,000 players and 2 million disks through the end of 1981 and went on to say that by 1990, production of videodisk players could reach five to six million per year, with sales of disks at 200 to 250 million.

Initial results don't bear out Philips's optimism. Magnavox players went on sale for the first time in Atlanta in December 1978 and at the time received an enthusiastic reception. But two of the Atlanta stores that introduced videodisk players report that sales plummeted in the next two years. Rich's Department Store sold only two players in the 1980 Christmas season, and a salesman said that some customers were returning videodisk players because of the lack of program material. Another Atlanta retailer said it sold about one videodisk player a month in 1980, far below the level of two years earlier. No official figures are available on overall sales, but there's one sure sign that sales are below expectations. Philips had stockpiled helium-neon lasers and established its own laser factory, because its projected need for helium-neon lasers to install in videodisk players was larger than the non-Communist world's entire manufacturing capability (about 150,000 lasers per year). But by early 1981, Philips had quietly begun selling some of the helium-neon lasers originally produced for its videodisk players to other companies.

Slow sales aren't the only problem. Many disks have turned out to be defective, and there have been some problems with players, which have been worsened by the shortage of technicians who know how to fix them. Even some of the dealers are discouraged. After three unsuccessful trips to a repair shop 20 miles (about 30 km) away, the operators of one Boston-area video store simply left their dead demonstration videodisk player in the middle of the store floor. When we wandered in to look at it, we were immediately told that it didn't work and were warned: "You know that it can't record."

RCA estimated that 26,000 of its videodisk players were sold to consumers during the first five weeks that they were on the market, a development claimed by the company to be "the most successful introduction of any major electronic product in history." The success was evidently news to some retailers, who began complaining about slow sales within a couple of months of the RCA players being introduced. Some dealers were said to be selling the players at their cost. It's not clear yet what, if any, technical problems the RCA players will experience. However, to be frank, some of RCA's marketing strategy appears ill-conceived, notably the efforts to market videodisks of musical performances with monophonic sound (stereo isn't yet available on the RCA player).

At best, the laser videodisk players are going to get a stiff run for their money. The competition comes not just from RCA and Matsushita, but from videotape recorders, the proliferation of multichannel cable-television systems, direct video broadcasting from satellites, and subscription television services, such as Home Box Office. The laser has given us the technology for a revolution in home entertainment. But it's uncertain if the laser, or any sort of videodisk system, is going to win that particular revolution.

VIDEODISK SPINOFFS

The technology developed for videodisk players is going to be around for a while, even if laser-based home videodisk players aren't. The special features available with laser videodisk players, such as the capability to stop the action, are particularly attractive for instructional use. The General Motors Corporation has already bought around 10,000 special "institutional" players, for use by its dealers in training and sales. So far, the dealers have yet to make extensive use of the videodisk training programs (although some bored salesmen made extensive use of the attached television sets during the sales slump of late 1980). But the technology appears promising for a variety of instructional uses, such as training employees for new tasks, where the same material must be presented over and over again at many locations.

A jukeboxlike arrangement of videodisks could serve as the basis of a video library, which could be tapped by the users of a fiber-optic cable-television network, such as we described in chapter 6. Another possibility is distributing some video programming in the form of videodisks to local cable-television systems, rather than transmitting it by satellites. This could be advantageous, because satellite transmission is vulnerable to unauthorized interception using increasingly available electronic equipment.

Philips has also adapted laser videodisk technology for an ultra-high-fidelity audio disk. The player, which is not yet on the market, uses a compact semiconductor laser, rather than a helium-neon gas laser, to play a small disk. Recording is digital rather than analog, giving a cleaner playback. At this writing, the compact digital disk appears well on its way to becoming accepted as a standard format for digital audio recordings. However, it's not clear when it will come on the market, or how large the market will be, especially given its incompatibility with existing records.

The same technology that can store video and audio information densely can also store digital information for retrieval by computers, as we mentioned briefly in chapter 11. So it looks as if the development of laser videodisk technology will have an impact—even if it's not on the home television screen, for which it was originally intended.

14 TOMORROW OR THE DAY AFTER: MORE FRONTIERS FOR THE LASER

We've mentioned many potential laser applications that have yet to mature, from high-energy laser weapons to low-energy laser communications over optical fibers. Those applications do not exhaust the laser's possibilities, however. Indeed, as we come to the close of this book, we feel twinges of guilt over some of the things we've left out. In this chapter, we'll touch on some of the other frontiers that laser technology may reach someday. Don't be surprised if you look back in a few years and see that we missed something. We'll be disappointed ourselves if scientists don't come up with some new and intriguing possibilities that no one today is even dreaming of.

GENETIC ENGINEERING

Genetic research is hampered by the tremendous difficulty in getting specific genes to change on demand. The basic problem is that genes are tiny and fragile, while the tools geneticists work with are crude and blunt. It's akin to trying to repair a watch with a sledge hammer. The fact that geneticists have accomplished as much as they have with such crude tools is a tribute to their ingenuity. A common approach is to use chemicals that randomly cause mutations (changes in genes) in large numbers of cells, then carefully screen the cells until the desired change is found. It's called shotgunning and, if done properly, is not quite as bad as looking for a needle in a haystack, since in this case the "needle" grows. Still, it can be tedious.

Geneticists would like to have a tool that could cause specific changes at specific points on a DNA molecule, and the laser is one candidate. The most obvious approach might seem to be to focus a laser beam so as to zap a tiny point on the DNA molecule. That turns out to be impossible, unfortunately, because the laser beam would have to be focused onto a point only about a billionth of a meter (4×10^{-8} in) in diameter. A practical laser with a wavelength short enough to accomplish this feat has yet to be developed. Light cannot be focused to a spot significantly smaller than its wavelength, and a billionth of a meter is roughly 1/500th the wavelength of visible light.

An alternative approach has been demonstrated by M. I. Stockmann of the Soviet Institute of Automation and Electrometry in Novosibirsk. He first added a dye, which attached itself to a specific part of the DNA molecule, then zapped the dye with pulses of ultraviolet light from a nitrogen laser. The dye absorbed the light, and enough energy was transferred to the DNA to break the genetic chain at that point. This technique, however, is still far from any practical application.

Other biological scientists are using lasers to study the details of photosynthesis, the process by which plants use sunlight to make carbohydrates. Photosynthesis is *fast;* it is measured in picoseconds (trillionths of a second), and lasers can produce pulses fast enough to "see" such rapid events. The basic idea is to shine a picosecond laser pulse on the photosynthetic apparatus of a cell, then watch what happens. The reality is much more complicated, because extremely complex instruments are needed to time the incredibly short events. The goal is to understand the now mysterious processes involved in photosynthesis—and perhaps ultimately to duplicate or improve on them.

GRAVITY WAVES

Gravity waves are tiny (and as yet undetected) oscillations that are thought to propagate through the very fabric of space–time at the speed of light. Albert Einstein predicted their existence in 1920, but it wasn't until 1960 that Joseph Weber of the University of Maryland (the same Joseph Weber who, while a graduate student at Columbia University, helped Charles Townes build the first maser) began a series of experiments intended to detect gravity waves. Weber built a huge metal bar, then looked for the minute mechanical distortions that, in theory, gravitational waves could cause in the bar. Other researchers have tried similar experiments, but at this writing there's no unambiguous evidence that gravity waves have been detected.

Because metal bar detectors haven't proved sensitive enough, physicists are looking at other types of gravity-wave detectors. One idea that's been studied by several groups relies on a fairly simple application of lasers. The basic idea is to suspend two massive mirrors in vacuum chambers that may be up to a few kilometers (or miles) or more apart. In theory, gravitational radiation should perturb the mirrors, causing them to move relative to one another. The resulting change in distance between the mirrors would still be tiny—somewhere around the wavelength of light or much less—even though the mirrors could be miles (or kilometers) away from each other. Even so, such minute displacements can be measured with laser interferometry, as described in the chapter on measurement (chapter 9). In fact, use of sophisticated signal-processing techniques allowed Robert L. Forward to measure periodic displacements as small as 9×10^{-16} m (roughly the diameter of a proton!) in 1972 experiments with a 8.5-m long (9.3 yd) gravity-wave detection system at Hughes Research Laboratories in Malibu, California. The qualification *periodic*—meaning that the displacements occur at regular intervals—is important, because such motion can be detected much more easily than motion that occurs at random intervals.

Although laser interferometry may seem sophisticated, it pales in comparison with an approach being studied by a group at the Soviet Institute of Thermophysics in Novosibirsk. S. N. Bagayev, V. G. Goldort, A. S. Dychkov, and V. P. Chebotayev have attained exquisite sensitivity by observing the motion of the mirrors that comprise the resonator of a very special type of laser—one that is stabilized to emit at a single, precisely defined wavelength. A minute change in the position of one mirror changes the length of the laser cavity, thereby causing a slight shift in the output wavelength. The shift in wavelength, in turn, causes a measurable change in the laser's output intensity. So far the Russians haven't found any gravity waves, but they have been able to detect periodic motion of as little as some 6×10^{-16} m—roughly the same as the best laser interferometer results.

There's still a long way to go. The longest existing laser interferometer for gravity-wave detection is a 10-m (11-yd) system under construction at the University of Glasgow, but researchers think they will need an interferometer a few kilometers (or miles) long to detect gravity waves. The Russians have only just begun testing their approach. There is another problem. Any instrument sensitive enough to detect gravity waves would also detect earth tremors and passing trucks, which means that physicists have to find a way to adjust the instrument so that these unwanted vibrations don't obscure the gravity waves they're looking for. Nonetheless, researchers are optimistic that sensitive laser-based measurements could someday verify Einstein's 60-year-old prediction.

PROVING SPECIAL RELATIVITY

A highly stabilized laser similar to the one used in the Russian gravity-wave experiments has also been used to test one of the most fundamental underpinnings of modern physics: Einstein's assumption that the speed of light is constant, no matter what direction it is traveling through space, an assumption essential to his theory of special relativity. At the Joint Institute for Laboratory Astrophysics in Boulder, Colorado, operated by the National Bureau of Standards and the University of Colorado, A. Brillet and John L. Hall performed experiments similar in intent to the famed Michelson-Morley experiment, which first showed that the speed of light is constant. The Brillet-Hall experiments were exquisitely sensitive. They demonstrated that there is no shift in the speed of light larger than about one part in 10^{15} (one quadrillion). Their results are the most precise test yet of Einstein's theory of special relativity.

Ultrastable lasers might also make it possible to define length in terms of time. The basic idea is to set a value for the speed of light (that is, to define the speed of light as equalling a certain number), then to define the international standard meter as the distance light travels in a certain fraction of a second. (All other units of length are defined in terms of the standard meter.) Such a redefinition isn't simply a concern of theorists. Defining the meter in terms of the speed of light would mean that instead of having to use the wavelength of light from a special type of lamp to define it, as is currently the case, physicists could use light from any sufficiently stable laser, according to Kenneth M. Evenson, a physicist at the National Bureau of Standards in Boulder. Evenson isn't alone; in 1981 the International Consultative Committee on the Definition of the Meter recommended that the speed of light be defined as equal to a certain value, and that the standard meter be defined in terms of the speed of light. At this writing, that recommendation had yet to be acted upon by international standards groups.

SUPER COOLERS

Physicists have also demonstrated some other exotic laser techniques, including one that permits cooling atoms or molecules to within a fraction of a degree of absolute zero—the temperature at which all motion stops, at least in theory. The basic idea is to tune a laser to a wavelength close to, but slightly different from, the wavelength of light normally absorbed by the atoms or molecules being studied, then shine it at the material. The natural motion of the atoms or molecules will shift their absorption wavelengths to a point where they'll absorb the laser light, which is then reemitted. Each time an atom or molecule absorbs and reemits light, it loses a tiny bit of energy—enough to cool it down slightly. If the laser light is intense enough, each atom or molecule absorbs and reemits light again and again, cooling off slightly each time. So far physicists have used this method to cool a few atoms to within a fraction of a degree of absolute zero. Such cold atoms are particularly interesting to scientists, because their low temperature simplifies their complex internal structure, making them easier to study than atoms at room temperature.

PHOTOCHEMISTRY

While there are major barriers to be overcome, lasers may one day be widely used to synthesize chemical compounds. Photons from a laser would excite atoms or molecules into states in which they would react readily with other atoms or molecules. Alternatively, laser photons might break up molecules or free electrons from atoms, rather than causing direct reactions with other atoms or molecules.

The idea of light causing chemical reactions is not in itself new. Sunlight causes many chemical reactions, ranging from sunburn to photosynthesis. Ordinary photography is based on a photochemical reaction: photons breaking down molecules of silver chloride to produce metallic silver (the black part of a photographic negative).

What's different about laser-induced chemical reactions is their potential precision. Earlier we described how each atom and molecule absorbs light at a unique set of wavelengths. To excite a particular atom or molecule, you need a laser that emits light at one of the wavelengths that the atom or molecule absorbs. Causing a chemical reaction is not quite that simple, though. You need to make certain that the light gives the atom or molecule enough energy to trigger such a reaction. You also need to make sure that the wavelength of the light you use isn't absorbed by other materials that the atoms or molecules of interest are mixed with. Finally, you need a way to extract the atoms or molecules produced by the chemical reactions.

So far most research on laser photochemistry has been focused on a problem we discussed in chapter 10 (on laser energy)—enriching the concentration of uranium-235 above the 0.7-percent level found in natural uranium. That's a particularly difficult problem, because the chemical properties of uranium-235 are, for all practical purposes, identical to those of the commoner uranium-238. However, the difficulties of conventional enrichment techniques are severe enough that, as we said earlier, the Department of Energy has maintained a large program using lasers to enrich the concentration of uranium-235.

Although many of the details of uranium-enrichment research remain classified, there has been widespread research on enrichment of isotopes of other elements, including hydrogen, chlorine, carbon, oxygen, sulfur, and osmium. Other researchers have studied more general properties of laser photochemistry. As a result, scientists have begun to understand what would be required for laser-induced chemistry to become practical on a large-scale basis. The results are somewhat encouraging.

The good news is that lasers can indeed produce chemical reactions in many types of materials. Demonstrations range from simple to elaborate: researchers at the Los Alamos National Laboratory, for example, have used a laser to ignite an alcohol lamp as well as to remove impurities from materials used to produce semiconductors for electronics.

The bad news is that laser techniques are often too expensive to compete with conventional alternatives, largely because lasers are relatively inefficient, which makes laser photons expensive. The problem isn't severe if the laser is inducing a chemical reaction involving only a small number of atoms or molecules, as is the case when removing impurities from semiconductor materials. But it becomes difficult to surmount when the goal is to excite almost every molecule in a mixture.

A further complication is that the types of lasers available often don't meet the needs of a particular job. The most noteworthy problem has been the search for a laser emitting light with a wavelength of 16 micrometers. That's the infrared wavelength at which uranium hexafluoride has a strong absorption that could be used, at least in theory, in isotope enrichment. Laser researchers, displaying their typical ingenuity, have come up with a long parade of sometimes bizarre lasers that emit in the required range, almost all of which have been both inefficient and cumbersome to operate. None have come anywhere near the efficiency and simplicity of the carbon dioxide laser, which emits at 10 micrometers. It was the coincidence of this 10-micrometer laser wavelength with a strong absorption of sulfur hexafluoride, a molecule somewhat similar to uranium hexafluoride, that stimulated much of the interest in laser enrichment of uranium in the first place. In the case of the 16-micrometer laser, it may be fortunate that a good one isn't right around the corner: it's not clear that the world needs a cheap and efficient way to enrich uranium-235, since that would remove the greatest impediment to easy production of atomic bombs.

Some of these problems will be overcome in the long run. Lasers will probably never be standard equipment in every chemical plant, but they may find work removing low concentrations of impurities where such impurities might prove harmful, as in semiconductor processing. Lasers may also prove useful in removing radioactive isotopes from low-level radioactive wastes, reducing the volume of dangerous wastes by concentrating the radioactive materials.

INTEGRATED OPTICS

A somewhat more futuristic technology is integrated optics. Today integrated electronic circuits perform many jobs that were once done by separate electronic components, such as transistors, resistors, and capacitors. What researchers would like to do is to integrate optical components into a single block of material in much the same way. Just as integrated electronics process electronic signals, so integrated optics would be used primarily to process optical signals and might also perform some computing functions that could be handled better optically than electronically.

The military, for instance, would like to use integrated optics to analyze radar signals and to simplify optical computing in satellites, missiles, and aircraft.

Integrated optics holds even greater potential in the rapid development of fiber-optic communications. At present, optical signals must be converted back into electronic signals for amplification, switching, or any sort of processing; then they must be reconverted into optical form for transmission through optical fibers. How much simpler it would be if the optical signal could be directly switched, processed, and amplified. Fiber-optics developers dream of all-optical networks, but at the moment it's only a dream.

LASER PROPULSION

At the opposite end of the power spectrum is a dream that's even further away from realization: laser propulsion of spacecraft and airliners. If it sounds crazy to you, you're not alone. The concept has raised the eyebrows of many physicists and engineers. But once you assume that high enough laser powers will become available—perhaps as an outgrowth of laser-weapon development—laser propulsion begins to look, if not easy, at least worth considering.

Distinctly different propulsion mechanisms are being studied for rockets and airplanes, though in both, the powering laser would be located not in the craft itself but on the ground or in a satellite. In a rocket, the laser beam would be directed against the bottom of the rocket, to heat a fuel of some sort. The heated fuel would boil off at extremely high speed. Exhausting the vaporized fuel downward (out of the back of the rocket) would push the rocket upward, just as if the fuel were being burned in an ordinary rocket. Various fuels are being considered, including hydrogen (because of its light weight), argon (because it's inert and readily available), and water (because it's also easily available and could be frozen).

In a laser-powered aircraft, the laser wouldn't heat a fuel, strictly speaking. Instead, it would heat air inside a jet engine, and the heated air would drive a turbine powering the airplane.

The two types of systems would also use vastly different types of lasers. Abraham Hertzberg of the University of Washington in Seattle and K. C. Sun of the Lockheed Palo Alto Research Laboratory in California have estimated that a solar power satellite equipped with a laser continuously delivering about 40 million watts would be enough to power an airplane. The beam would be directed, perhaps via relay satellite, at the top of an aircraft while it was at its cruising altitude. To get from the ground to cruising altitude and back down again, the plane would use standard aviation fuel—kerosene.

More than ten times as much laser power—ranging up to over a hundred times as much for a heavy payload—would be needed to get a laser-powered rocket off the ground. The reason should be obvious if you've ever watched a rocket launch. A tremendous amount of thrust is needed to get a rocket going. In practice a launch laser would probably be located on top of a mountain—not to gain a few kilometers (or miles) in height but to get above as much of the atmosphere as possible. As we saw earlier, the atmosphere can play havoc with a laser beam, and compensating for atmospheric effects so that almost the entire power of the launch laser can be focused onto the bottom of the rocket is considered a major problem. However, it is not as serious as the problem of aiming laser weapons, because the rocket would be a "cooperative" target, which could be designed to assist in focusing the laser beam.

NASA researchers have concluded that ground-based laser propulsion isn't in the near future because of the tremendous amounts of laser power that would be required. However, laser propulsion could also be used to shift spacecraft between orbits—a task that will become particularly important once the Space Shuttle is routinely putting spacecraft into low earth orbit. A laser rocket is expected to require only about 10 percent as much propellant as a chemical rocket. Because it's very expensive to put fuel into orbit, that weight reduction could result in big savings, even after building the massive laser that would be required. Operating in space would require laser powers of several tens of megawatts (much less than for a ground-based laser) while also avoiding the atmospheric effects that could present problems for a ground-based system.

These proposals all depend on our being able to build BIG lasers. A 40-million-watt laser that might power an airplane or shift satellites between orbits would be nearly ten times as powerful as the first-generation space-based laser weapons that have been proposed by Senator Wallop. Unlike weapons, propulsion lasers would have to operate continuously or produce a series of very rapid pulses for periods of hours. A ground-based launch laser for a rocket would have to supply an average power of about a *billion* watts over a period of more than 30 minutes. Building such lasers is probably not impossible, but it's clearly going to be a challenge.

Why, then, are NASA and DOD interested in laser propulsion? In theory, it turns out that laser propulsion could be much cheaper and cleaner than present techniques. Laser propulsion also promises to offer higher thrust in space than other efficient propulsion schemes such as ion or electric drives.

Hertzberg and Sun came up with some interesting numbers to support their ideas for laser-propelled airplanes. One of their proposed aircraft would need only about 4,800 kilograms (5 tons) of kerosene for a 5,500-km (3,500-mile) transcontinental flight, whereas a conventional airplane with a new, fuel-saving design would burn about 29,500 kilograms (33 tons) of kerosene for the same flight. The capital costs would be high—an estimated $100 million each for 300 solar power satellites and about $50 million each for 400 relay satellites. In 1978 Hertzberg and Sun calculated that their approach would be no more expensive than conventional transcontinental flight, should the price of aviation fuel reach $1.40 per gallon—a price that may already be here by the time you read this. What's more, the energy would come from the sun, not from our declining petroleum supplies.

Dramatic economies are also predicted for laser-driven rockets. Arthur Kantrowitz, who suggested the idea of laser propulsion over a decade ago, while he was head of the Avco Everett Research Laboratory in Everett, Massachusetts, has talked about putting satellites into orbit for about $20 per pound ($40 per kilogram), compared with some $1,000 per pound ($2,000 per kilogram), for single-shot rockets and about half that using the Space Shuttle. A Lockheed study predicts that laser rockets for orbital transfer should be several times less expensive than chemical rockets.

Economics isn't the only concern. Advocates of intensive space development aren't blind to the pollution problems created by present-day chemical rockets, because pollution can put an upper limit on how many launches are feasible. Laser-driven rockets could use nonpolluting propellants, such as water or argon, and entirely avoid the problem. (The only hitch would be a chemical type of ground-based laser that could emit chemical pollutants.) Kantrowitz has talked about a laser system that could launch 1 ton into orbit every few minutes, putting some 100,000 tons of freight into orbit per year.

Laser propulsion still has some physicists shaking their heads in amazement that anyone could take it seriously—and with good reason.

It is far beyond the current state of the art. Yet such bold ideas can serve as important stimuli to our technological imaginations, even if they are never realized in their original form.

CONVERSATIONS WITH OTHER WORLDS

There are even more exotic possibilities than laser propulsion. Lasers are also candidates for interstellar communications devices.

Pictures of the planets are sent back to earth by spacecraft such as Voyager via radio waves or microwaves. Edward C. Posner of the California Institute of Technology has a plan for using a laser to transmit such information over vastly longer distances—distances that radio waves or microwaves may not be able to handle. The power required would be surprisingly modest. Posner has calculated that a laser beam with an average power of only 500 watts could transmit a planetary image, similar in quality to the Voyager images of the outer planets, to the earth from a distance of six light-years. Specifically, he was considering exploration of the Barnard's Star system, six light-years away from the earth, which is considered likely to have planets.

Posner's estimate is somewhat optimistic. He assumes a sophisticated coding scheme and assumes that light from the star would not obscure light from the laser on the interstellar probe. Those assumptions may be too optimistic, but they point out some of the capabilities of laser transmission.

In principle, with high enough laser powers and narrow enough beams, it should even be possible to transmit signals across the width of the galaxy, although in practice they would probably be obscured by absorption by interstellar dust. The laser could be pumped by energy from a star, and a large space-borne mirror—somewhere around a kilometer (mile) across—could be used to focus the beam.

Building the hardware would probably be the easiest part of setting up such a communications link. The hardest part would be establishing communications in the first place. The extremely narrow beam from the laser simply wouldn't be suitable for searching out other civilizations; it would almost certainly miss any that were out there. Once a target was found, aiming the beam would be no minor task either, since it would require a knowledge of motion of the target star system for the 100,000 years it would take the laser beam to get across the galaxy. (Over that distance the beam would spread out to only about the diameter of the solar system.) It's the type of problem that makes building a laser propulsion system seem simple by comparison.

OTHER FUTURISTIC APPLICATIONS

Other futuristic ideas make more—or less—sense than interstellar communications. Several years ago a group in Texas bought a laser with which to try to communicate with unidentified flying objects (UFOs). The fact that we haven't heard from the group since may indicate that they haven't heard from any UFOs. A group of Chinese researchers have found that hair grows faster on skin illuminated by laser light, but so far we haven't heard any reports of using laser light to help grow hair on bald heads. There have been serious proposals to use high-energy lasers to beam power either from point to point on the earth via relay satellites or from a master power satellite to other satellites that would derive their energy from the master satellite. And it's a fairly safe bet that someone in a basement or back room somewhere is trying to use a laser to contact the spirit world.

THE NEXT TWENTY YEARS

The next twenty years of laser development may bring only a handful of these ideas to fruition. Some of today's major efforts may be relegated to the footnotes of interesting, but impractical, ideas. Others will still be in the development stages, probably far behind the earliest optimistic projections but still showing promise. And a few will have yielded the laser tools of the twenty-first century.

If you want to know which efforts will work and which won't, you'll have to wait. That's what the research-and-development process is all about. It's not supposed to turn every single idea into something of earth-shaking significance, but rather to test out the potential of each idea, and to see what ideas solve real problems at a more reasonable cost than other alternatives. Few of the laser scientists we talked to were willing to go out on a limb without seeing the results, and we'll stick with them.

The past two decades have seen the transformation of the laser from the embodiment of one of the most persistent visions of science fiction to a practical tool for an increasing variety of uses. The number of ways in which lasers are used and the number of lasers being used are almost certain to increase over the next two decades.

The laser will be more likely than ever to come into your life.

APPENDIX: LASER SAFETY

Lasers are not the fearsome instruments of destruction science-fiction writers make them out to be. And, in fact, many lasers are used to help cure disease. But incorrect use of a laser can cause injury.

From a practical standpoint, the biggest danger is not the laser beam but the high voltage required to produce it. Laser power supplies can generate the combination of voltage and current needed to kill someone, and three or four researchers have been electrocuted. This danger is not special to lasers; it's present with all high-voltage equipment.

Careless exposure to light from the types of lasers you're likely to encounter can at worst cause eye damage. Although there is little energy in a low-power laser beam, that power is concentrated on a very small spot. The intensity of a beam from a low-power laser, such as you might find in a high-school physics laboratory, is about the same as the intensity of sunlight. What's more, the light rays in a laser beam are parallel. That makes it a potential hazard to the eye, because those intense, parallel rays could be focused by the eye to an extremely small point. Just as it's dangerous to stare into the sun, it's dangerous to stare into the beam from a laser. Momentary exposure shouldn't be hazardous; the real danger is letting a laser beam stay focused on a single point on the retina.

Although real, the hazards are not great. In the twenty years since the development of the laser, only about twenty eye accidents have been reported to the federal government. In some of these, no permanent damage was found. Virtually all of these accidents could have been prevented by such rudimentary precautions as wearing special safety goggles that selectively block light at laser wavelengths.

There are both voluntary standards and government-mandated regulations designed to promote laser safety. The most important of the voluntary standards in the United States are those issued by the American National Standards Institute. This is a codification of procedures recommended for safe use of lasers in laboratories and industry.

The Occupational Safety and Health Administration (OSHA) has taken no action since issuing a *proposed* standard several years ago. This standard has never taken effect, and though OSHA is the only federal agency with the authority to set mandatory rules for the *use* of lasers, it has taken no action, apparently believing that it has more important things to worry about than laser safety.

The main federal agency involved in setting rules for laser safety is the Bureau of Radiological Health (BRH). BRH has the authority to establish safety rules that must be met by the makers of laser hardware but does not have the authority to set rules on the use of lasers. Many people in the laser industry feel that BRH's rules are unnecessarily strict and require too much paperwork, thereby driving up the cost of laser equipment and sometimes preventing the use of lasers. One of the largest companies in the United States is reported to have told its engineers to minimize the use of lasers so as to avoid paperwork requirements.

State regulations on the use of lasers range from nonexistent to stringent. New York State's requirements are generally considered to be the most severe; they include mandatory licensing of the operators of even small helium-neon lasers. BRH is trying to coordinate development of a standard state code, but at this writing that effort has made little progress.

A CATALOG OF SELECTED

DOVER BOOKS

IN ALL FIELDS OF INTEREST

FRANK LLOYD WRIGHT'S HOLLYHOCK HOUSE, Donald Hoffmann. Lavishly illustrated, carefully documented study of one of Wright's most controversial residential designs. Over 120 photographs, floor plans, elevations, etc. Detailed perceptive text by noted Wright scholar. Index. 128pp. 9¼ x 10¾. 27133-1 Pa. $11.95

THE MALE AND FEMALE FIGURE IN MOTION: 60 Classic Photographic Sequences, Eadweard Muybridge. 60 true-action photographs of men and women walking, running, climbing, bending, turning, etc., reproduced from rare 19th-century masterpiece. vi + 121pp. 9 x 12. 24745-7 Pa. $10.95

1001 QUESTIONS ANSWERED ABOUT THE SEASHORE, N. J. Berrill and Jacquelyn Berrill. Queries answered about dolphins, sea snails, sponges, starfish, fishes, shore birds, many others. Covers appearance, breeding, growth, feeding, much more. 305pp. 5¼ x 8¼. 23366-9 Pa. $8.95

GUIDE TO OWL WATCHING IN NORTH AMERICA, Donald S. Heintzelman. Superb guide offers complete data and descriptions of 19 species: barn owl, screech owl, snowy owl, many more. Expert coverage of owl-watching equipment, conservation, migrations and invasions, etc. Guide to observing sites. 84 illustrations. xiii + 193pp. 5⅜ x 8½. 27344-X Pa. $8.95

MEDICINAL AND OTHER USES OF NORTH AMERICAN PLANTS: A Historical Survey with Special Reference to the Eastern Indian Tribes, Charlotte Erichsen-Brown. Chronological historical citations document 500 years of usage of plants, trees, shrubs native to eastern Canada, northeastern U.S. Also complete identifying information. 343 illustrations. 544pp. 6½ x 9¼. 25951-X Pa. $12.95

STORYBOOK MAZES, Dave Phillips. 23 stories and mazes on two-page spreads: Wizard of Oz, Treasure Island, Robin Hood, etc. Solutions. 64pp. 8¼ x 11. 23628-5 Pa. $2.95

NEGRO FOLK MUSIC, U.S.A., Harold Courlander. Noted folklorist's scholarly yet readable analysis of rich and varied musical tradition. Includes authentic versions of over 40 folk songs. Valuable bibliography and discography. xi + 324pp. 5⅜ x 8½. 27350-4 Pa. $9.95

MOVIE-STAR PORTRAITS OF THE FORTIES, John Kobal (ed.). 163 glamor, studio photos of 106 stars of the 1940s: Rita Hayworth, Ava Gardner, Marlon Brando, Clark Gable, many more. 176pp. 8⅜ x 11¼. 23546-7 Pa. $12.95

BENCHLEY LOST AND FOUND, Robert Benchley. Finest humor from early 30s, about pet peeves, child psychologists, post office and others. Mostly unavailable elsewhere. 73 illustrations by Peter Arno and others. 183pp. 5⅜ x 8½. 22410-4 Pa. $6.95

YEKL and THE IMPORTED BRIDEGROOM AND OTHER STORIES OF YIDDISH NEW YORK, Abraham Cahan. Film Hester Street based on Yekl (1896). Novel, other stories among first about Jewish immigrants on N.Y.'s East Side. 240pp. 5⅜ x 8½. 22427-9 Pa. $6.95

SELECTED POEMS, Walt Whitman. Generous sampling from *Leaves of Grass*. Twenty-four poems include "I Hear America Singing," "Song of the Open Road," "I Sing the Body Electric," "When Lilacs Last in the Dooryard Bloom'd," "O Captain! My Captain!"–all reprinted from an authoritative edition. Lists of titles and first lines. 128pp. 5$\frac{5}{16}$ x 8¼. 26878-0 Pa. $1.00

MY BONDAGE AND MY FREEDOM, Frederick Douglass. Born a slave, Douglass became outspoken force in antislavery movement. The best of Douglass' autobiographies. Graphic description of slave life. 464pp. 5⅜ x 8½. 22457-0 Pa. $8.95

FOLLOWING THE EQUATOR: A Journey Around the World, Mark Twain. Fascinating humorous account of 1897 voyage to Hawaii, Australia, India, New Zealand, etc. Ironic, bemused reports on peoples, customs, climate, flora and fauna, politics, much more. 197 illustrations. 720pp. 5⅜ x 8½. 26113-1 Pa. $15.95

THE PEOPLE CALLED SHAKERS, Edward D. Andrews. Definitive study of Shakers: origins, beliefs, practices, dances, social organization, furniture and crafts, etc. 33 illustrations. 351pp. 5⅜ x 8½. 21081-2 Pa. $8.95

THE MYTHS OF GREECE AND ROME, H. A. Guerber. A classic of mythology, generously illustrated, long prized for its simple, graphic, accurate retelling of the principal myths of Greece and Rome, and for its commentary on their origins and significance. With 64 illustrations by Michelangelo, Raphael, Titian, Rubens, Canova, Bernini and others. 480pp. 5⅜ x 8½. 27584-1 Pa. $9.95

PSYCHOLOGY OF MUSIC, Carl E. Seashore. Classic work discusses music as a medium from psychological viewpoint. Clear treatment of physical acoustics, auditory apparatus, sound perception, development of musical skills, nature of musical feeling, host of other topics. 88 figures. 408pp. 5⅜ x 8½. 21851-1 Pa. $10.95

THE PHILOSOPHY OF HISTORY, Georg W. Hegel. Great classic of Western thought develops concept that history is not chance but rational process, the evolution of freedom. 457pp. 5⅜ x 8½. 20112-0 Pa. $9.95

THE BOOK OF TEA, Kakuzo Okakura. Minor classic of the Orient: entertaining, charming explanation, interpretation of traditional Japanese culture in terms of tea ceremony. 94pp. 5⅜ x 8½. 20070-1 Pa. $3.95

LIFE IN ANCIENT EGYPT, Adolf Erman. Fullest, most thorough, detailed older account with much not in more recent books, domestic life, religion, magic, medicine, commerce, much more. Many illustrations reproduce tomb paintings, carvings, hieroglyphs, etc. 597pp. 5⅜ x 8½. 22632-8 Pa. $11.95

SUNDIALS, Their Theory and Construction, Albert Waugh. Far and away the best, most thorough coverage of ideas, mathematics concerned, types, construction, adjusting anywhere. Simple, nontechnical treatment allows even children to build several of these dials. Over 100 illustrations. 230pp. 5⅜ x 8½. 22947-5 Pa. $7.95

DYNAMICS OF FLUIDS IN POROUS MEDIA, Jacob Bear. For advanced students of ground water hydrology, soil mechanics and physics, drainage and irrigation engineering, and more. 335 illustrations. Exercises, with answers. 784pp. 6⅛ x 9¼. 65675-6 Pa. $19.95

SONGS OF EXPERIENCE: Facsimile Reproduction with 26 Plates in Full Color, William Blake. 26 full-color plates from a rare 1826 edition. Includes "The Tyger," "London," "Holy Thursday," and other poems. Printed text of poems. 48pp. 5¼ x 7. 24636-1 Pa. $4.95

OLD-TIME VIGNETTES IN FULL COLOR, Carol Belanger Grafton (ed.). Over 390 charming, often sentimental illustrations, selected from archives of Victorian graphics–pretty women posing, children playing, food, flowers, kittens and puppies, smiling cherubs, birds and butterflies, much more. All copyright-free. 48pp. 9¼ x 12¼. 27269-9 Pa. $7.95

EARLY NINETEENTH-CENTURY CRAFTS AND TRADES, Peter Stockham (ed.). Extremely rare 1807 volume describes to youngsters the crafts and trades of the day: brickmaker, weaver, dressmaker, bookbinder, ropemaker, saddler, many more. Quaint prose, charming illustrations for each craft. 20 black-and-white line illustrations. 192pp. 4⅝ x 6. 27293-1 Pa. $4.95

VICTORIAN FASHIONS AND COSTUMES FROM HARPER'S BAZAR, 1867–1898, Stella Blum (ed.). Day costumes, evening wear, sports clothes, shoes, hats, other accessories in over 1,000 detailed engravings. 320pp. 9⅜ x 12¼. 22990-4 Pa. $14.95

GUSTAV STICKLEY, THE CRAFTSMAN, Mary Ann Smith. Superb study surveys broad scope of Stickley's achievement, especially in architecture. Design philosophy, rise and fall of the Craftsman empire, descriptions and floor plans for many Craftsman houses, more. 86 black-and-white halftones. 31 line illustrations. Introduction 208pp. 6½ x 9¼. 27210-9 Pa. $9.95

THE LONG ISLAND RAIL ROAD IN EARLY PHOTOGRAPHS, Ron Ziel. Over 220 rare photos, informative text document origin (1844) and development of rail service on Long Island. Vintage views of early trains, locomotives, stations, passengers, crews, much more. Captions. 8⅞ x 11¾. 26301-0 Pa. $13.95

THE BOOK OF OLD SHIPS: From Egyptian Galleys to Clipper Ships, Henry B. Culver. Superb, authoritative history of sailing vessels, with 80 magnificent line illustrations. Galley, bark, caravel, longship, whaler, many more. Detailed, informative text on each vessel by noted naval historian. Introduction. 256pp. 5⅜ x 8½. 27332-6 Pa. $7.95

TEN BOOKS ON ARCHITECTURE, Vitruvius. The most important book ever written on architecture. Early Roman aesthetics, technology, classical orders, site selection, all other aspects. Morgan translation. 331pp. 5⅜ x 8½. 20645-9 Pa. $8.95

THE HUMAN FIGURE IN MOTION, Eadweard Muybridge. More than 4,500 stopped-action photos, in action series, showing undraped men, women, children jumping, lying down, throwing, sitting, wrestling, carrying, etc. 390pp. 7⅞ x 10⅝. 20204-6 Clothbd. $25.95

TREES OF THE EASTERN AND CENTRAL UNITED STATES AND CANADA, William M. Harlow. Best one-volume guide to 140 trees. Full descriptions, woodlore, range, etc. Over 600 illustrations. Handy size. 288pp. 4½ x 6⅜. 20395-6 Pa. $6.95

SONGS OF WESTERN BIRDS, Dr. Donald J. Borror. Complete song and call repertoire of 60 western species, including flycatchers, juncoes, cactus wrens, many more–includes fully illustrated booklet. Cassette and manual 99913-0 $8.95

GROWING AND USING HERBS AND SPICES, Milo Miloradovich. Versatile handbook provides all the information needed for cultivation and use of all the herbs and spices available in North America. 4 illustrations. Index. Glossary. 236pp. 5⅜ x 8½. 25058-X Pa. $6.95

BIG BOOK OF MAZES AND LABYRINTHS, Walter Shepherd. 50 mazes and labyrinths in all–classical, solid, ripple, and more–in one great volume. Perfect inexpensive puzzler for clever youngsters. Full solutions. 112pp. 8⅛ x 11. 22951-3 Pa. $4.95